Cambridge IGCSE®
Chemistry

Practical Workbook

Michael Strachan

CAMBRIDGE
UNIVERSITY PRESS

University Printing House, Cambridge CB2 8BS, United Kingdom

Cambridge University Press is part of the University of Cambridge.

It furthers the University's mission by disseminating knowledge in the pursuit of education, learning and research at the highest international levels of excellence.

Information on this title: education.cambridge.org

© Cambridge University Press 2016

This publication is in copyright. Subject to statutory exception and to the provisions of relevant collective licensing agreements, no reproduction of any part may take place without the written permission of Cambridge University Press.

First published 2016

Printed in the United Kingdom by Latimer Trend

A catalogue record for this publication is available from the British Library

ISBN 978-1-316-60946-0 Paperback

The questions, answers and annotation in this title were written by the author and have not been produced by Cambridge International Examinations.

Cambridge University Press has no responsibility for the persistence or accuracy of URLs for external or third-party internet websites referred to in this publication, and does not guarantee that any content on such websites is, or will remain, accurate or appropriate. Information regarding prices, travel timetables, and other factual information given in this work is correct at the time of first printing but Cambridge University Press does not guarantee the accuracy of such information thereafter.

....

NOTICE TO TEACHERS IN THE UK

It is illegal to reproduce any part of this work in material form (including photocopying and electronic storage) except under the following circumstances:
(i) where you are abiding by a licence granted to your school or institution by the Copyright Licensing Agency;
(ii) where no such licence exists, or where you wish to exceed the terms of a licence, and you have gained the written permission of Cambridge University Press;
(iii) where you are allowed to reproduce without permission under the provisions of Chapter 3 of the Copyright, Designs and Patents Act 1988, which covers, for example, the reproduction of short passages within certain types of educational anthology and reproduction for the purposes of setting examination questions.

....

IGCSE ® is the registered trademark of Cambridge International Examinations

Contents

Introduction		v
Safety section		vi
Skills grid		vii
Quick skills section		ix

1 Planet Earth 1
   1.1 Estimating the percentage of oxygen in the air   1
   1.2 The effects of acid rain on metal   5
   1.3 The effect of carbon dioxide on the atmosphere   9

2 The nature of matter 16
   2.1 Changing physical state   16
   2.2 Filtration, distillation and evaporation   20
   2.3 Chromatography   24

3 Elements and compounds 32
   3.1 The properties of metals and non-metals   32
   3.2 The differences between elements and compounds   37
   3.3 The properties of ionic and covalent compounds   39

4 Chemical reactions 47
   4.1 Types of chemical reaction   47
   4.2 Exothermic and endothermic reactions   51
   4.3 The electrolysis of copper   54

5 Acids, bases and salts 61
   5.1 Weak and strong acids   61
   5.2 Reacting acids   65
   5.3 Reacting alkalis   68
   5.4 The preparation of soluble salts   71
   5.5 The preparation of insoluble salts   74
   5.6 Acid titration   76
   5.7 The pH of oxides   79

6 Quantitative chemistry 84
   6.1 Determination of the relative atomic mass of magnesium   84
   6.2 Calculating the empirical formula of hydrated salts   87
   6.3 Calculating percentage yield using copper carbonate   90
   6.4 Calculating the concentration of acid using the titration method   93

7 How far? How fast? 99
   7.1 The effect of temperature on reaction rate   99
   7.2 The effect of catalysts on reaction rate   103
   7.3 Energy changes during displacement reactions   107
   7.4 Reversible reactions   111

8 Patterns and properties of metals 116
   8.1 The extraction of iron   116
   8.2 The extraction of copper from malachite   118
   8.3 The reactions of metals and acids   120
   8.4 Investigating the reactivity series using an electrochemical cell   123
   8.5 Using flame tests to identify metals   126

9 Industrial inorganic chemistry 133
   9.1 What causes rusting?   133
   9.2 Preventing rusting   137
   9.3 Using limestone (calcium carbonate) to neutralise acidic water   140

10 Organic chemistry 146
   10.1 Testing for alkanes and alkenes   146
   10.2 Fermentation of glucose using yeast   149
   10.3 Making esters from alcohols and acids   151

11 Petrochemicals and polymers 155
   11.1 Cracking hydrocarbons   155
   11.2 Comparing polymers   159
   11.3 Comparing fuels   161

12 Chemical analysis and investigation 167
   12.1 Identifying anions   167
   12.2 Identifying cations   171

Introduction

None of the pioneering work done in the scientific field of chemistry would ever have occurred without the laboratory and the art of experimentation. Nearly all of the great discoveries that form the foundations of our knowledge came from people completing practical investigations in laboratories not too different from the ones you will be using to complete your studies. Great chemists such as Lavoisier and Priestley would recognise some of the principles contained in this book, though they would note the modern approaches used to investigate them.

Practical skills form the backbone of any chemistry course and it is hoped that, by using this book, you will gain confidence in this exciting and essential area of study. This book has been written for the Cambridge IGCSE Chemistry student. The various investigations and accompanying questions will help you to build and refine your abilities. You will gain enthusiasm in tackling laboratory work and will learn to demonstrate a wide range of practical skills. Aside from the necessary preparation for both the practical paper and the alternative to practical paper, it is hoped that these interesting and enjoyable investigations will kindle a deep love of practical chemistry in you. Great care has been taken to ensure that this book contains work that is safe and accessible for you to complete. Before attempting any of these activities, make sure that you have read the safety section and are following the safety regulations of the place where you study. Answers to the exercises in this Workbook can be found in the Teacher's guide. Ask your teacher to provide access to the answers.

Safety section

Despite using Bunsen burners and chemicals on a regular basis, the science laboratory is one of the safest classrooms in a school. This is due to the emphasis on safety and the following of precautions set out by regular risk assessment and procedures.

It is important that you follow the safety rules set out by your teacher. Your teacher will know the names of materials and the hazards associated with them as part of their risk assessment for performing the investigations. They will share this information with you as part of their safety briefing or demonstration of the investigation.

The safety precautions in each of the investigations of this book are guidance that you should follow to ensure your safety and that of other students around you. You should aim to use the safety rules as further direction to help to prepare for examination when planning your own investigations in the alternative to practical papers.

The following precautions will help to ensure your safety when carrying out most investigations in this workbook.

- Wear safety goggles to protect your eyes.
- Tie back hair and any loose items of clothing.
- Tidy away personal belongings to avoid tripping over them.
- Wear gloves and protective clothing as described by the book or your teacher.
- Turn the Bunsen burner to the cool, yellow flame when not in use.
- Observe hazard symbols and chemical information provided with all substances and solutions.

Many of the investigations require some sort of teamwork or group work. It is the responsibility of your group to make sure that you plan how to be safe as diligently as you plan the rest of the investigation.

In Chemistry particular attention should be paid to the types of Bunsen burner flame needed as well as the concentrations and volumes of chemicals used.

Skills grid

Assessment objective 3 (AO3) 'Experimental skills and investigations' of the Cambridge International Examinations syllabus is about your ability to work as a scientist. Each aspect of the AO3 has been broken down for you below with a reference to the chapters in this title that cover it. This will enable you to identify where you have practiced each skill and also allow you to revise each one before the exam.

Chapter	1	2	3	4	5	6	7	8	9	10	11	12
AO3: Experimental skills and investigations												
1.1 demonstrate knowledge of how to safely use techniques		X	X	X	X	X		X		X	X	
1.2 demonstrate knowledge of how to use apparatus and materials		X	X	X	X	X	X	X		X	X	
1.3 demonstrate knowledge of how to follow a sequence of instructions where appropriate	X	X	X	X	X	X	X	X	X	X	X	X
2. plan experiments and investigations		X	X					X	X	X	X	
3.1 make and record observations			X	X	X	X	X	X	X	X	X	X
3.2 make and record measurements	X	X	X	X	X	X	X	X			X	
3.3 make and record estimates		X		X	X		X					
4.1 interpret experimental observations and data	X	X	X	X	X	X	X	X	X	X	X	X
4.2 evaluate experimental observations and data				X	X	X	X	X		X		
5.1 evaluate methods	X			X	X	X	X	X				
5.2 suggest possible improvements to methods	X	X	X	X	X	X			X		X	

Quick skills section

Apparatus

You will need to be able to identify, use and draw a variety of scientific apparatus. Complete the table below by adding the diagram and uses for each piece of apparatus. The first two have been completed for you.

Apparatus	Diagram	Uses
timer		To measure the time taken for something to happen. Usually measured in seconds.
balance		To measure the mass of a substance. Usually measured in grams.
thermometer		
measuring cylinder		
beaker		
pipette		
burette		
conical flask		
Bunsen burner		
tripod		
test-tube / boiling tube		

Measuring

Being able to take accurate measurements is an essential skill for all chemistry students. As part of the Cambridge IGCSE course you will be expected to be able to take accurate measurements using a variety of different apparatus. When using measuring cylinders you will need to look for the meniscus, which is the bottom of the curve formed by the liquid.

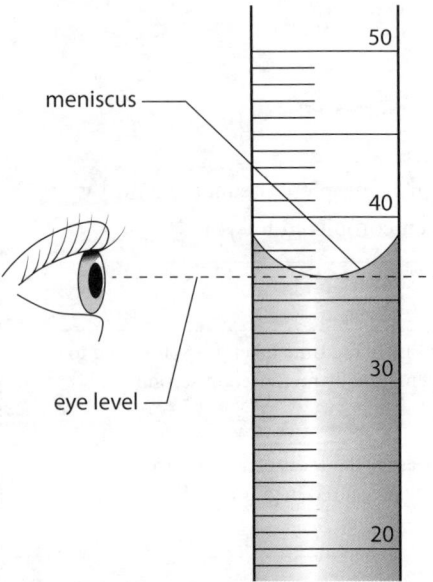

Thermometers are a very common tool for measuring temperature in chemistry experiments so you will need to be able to take readings reliably. Not all of the points of the scale on a thermometer will be marked but you will still need to be able to determine the temperature. To do this you will need to work out the value of each graduation. In the diagram below there are four marks between 95 and 100. Each of these marks indicates 1°C.

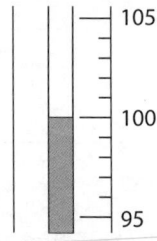

Recording

When working on investigations the ability to record data accurately is very important. Sometimes a table will be supplied; however, you need to be able to draw your own table with the correct headings and units.
The first task is to identify the independent and dependent variables for the investigation you are doing. The independent variable is the one that you are changing to see if this affects the dependent variable. The dependent variable is the one that you will measure and record the results of in the table. The names of these two variables and their units need to go into the top two boxes in your results table. The independent variable goes in the left hand box and the dependent variable goes in the right hand box. Separate the name of the variables and units using a forward slash, e.g. time / seconds. Remember that the column headings need to be physical quantities (time, mass, temperature, etc.).

Next count how many different values you have for the independent variable. This is how many rows you will need to add below the column headings. Finally add the values for the independent variable into the left hand column. Your table is now ready for you to add the results from your investigation into the right hand column.

Independent variable / units	Dependent variable / units

The number of significant figures that you use in your answer should match the number used in any data you have been given. You may not be awarded credit if you use an inappropriate number of significant figures in your answers to questions. If you are recording raw data from an investigation, always try to use the maximum number of significant figures available.

The first significant figure is the first non-zero digit in the number. The number 456 is: 500 to 1 significant figure; 460 to 2 significant figures; 456 to 3 significant figures; 456.0 to 4 significant figures, etc. Digits of 5 or greater are rounded up; and digits of 4 and below are rounded down. It is important that numbers are not rounded up during calculations until you have your final answer, otherwise the final answer may be affected.

Graphing

When drawing a graph it is useful to follow a set procedure every time to ensure that when you are finished the graph is complete.

Axes: You must label the axes with your independent and dependent variables. The independent variable is used to label the *x*-axis (horizontal axis) and the dependent variable is used to label the *y*-axis (vertical axis). Remember to also add the units for each of the variables. An easy way to ensure that you get this correct is to copy the column headings from the table of data you are using to draw the graph.

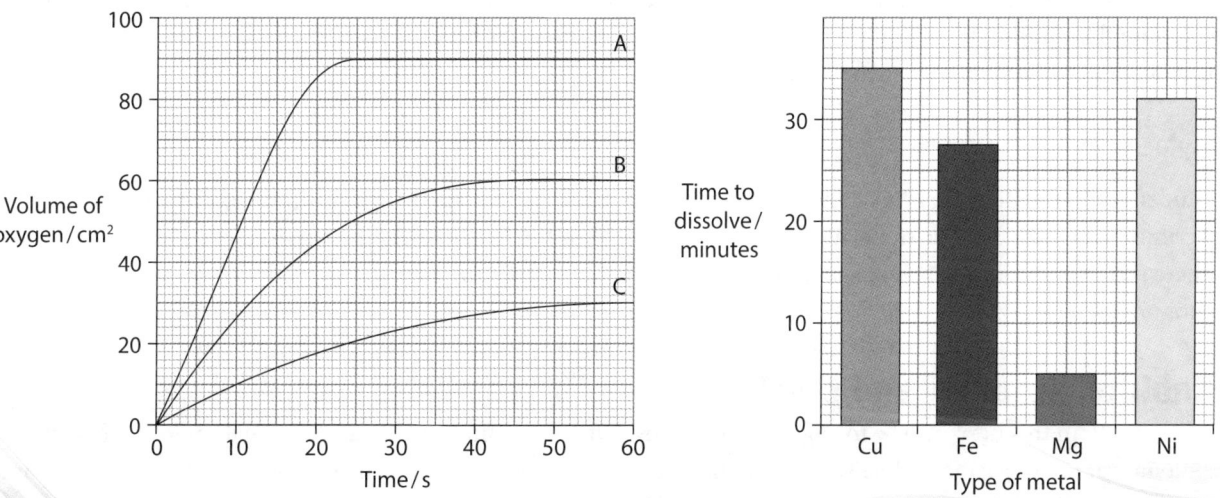

Tip: At the top of any table of data you have to use write the letters X and Y next to the independent and dependent variable to remind you which axis each goes on.

The second stage of drawing a graph is adding a scale. You must select a scale that allows you to use more than half of the graph grid in both directions. Choose a sensible ratio to allow you to easily plot your points (e.g. each 1cm on the graph grid represents 1, 2, 5, 10, 50 or 100 units of the variable). If you choose to use other numbers for your scale it becomes much more difficult to plot your graph.

Now you are ready to plot the points of data on the graph grid. You can use either crosses (x) or a point enclosed inside a circle (•) to plot your points but take your time to make sure these are plotted accurately. Remember to use a sharp pencil as large dots make it difficult to see the place the point is plotted and may make it difficult for the accuracy of the plot to be decided.

Finally a best-fit line needs to be added. This must be a single thin line or smooth curve. It does not need to go through all of the points but it should have roughly half the number of points on each side of the line or curve. Remember to ignore any anomalous data when you draw your best-fit line. Some good examples of best-fit lines that you should use are shown below:

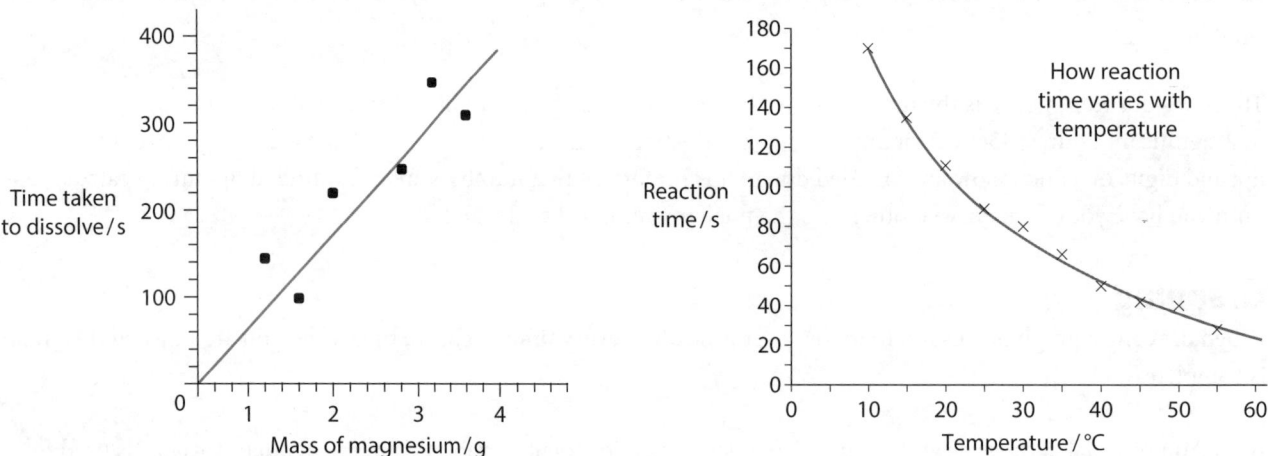

Variables

The independent and dependent variables have already been discussed but there is a third type of variable that you will need to be familiar with – controlled variables. These are variables that are kept the same during an investigation to make sure that they do not affect the results. If these variables are not kept the same then we cannot be sure that it is our independent variable having an effect on the results.

Example

Two students are investigating how changing the temperature affects the rate that gas is produced when adding magnesium to an acid. They do not control the volume of acid or the mass of magnesium used each time. This means that there is no pattern in their results, as if they use more acid magnesium more gas is produced regardless of the temperature used.

Reliability, accuracy and precision

A common task in this book will be to suggest how to improve the method used in an investigation to improve its reliability/accuracy/precision. Before we come to how these improvements can be made it is important that you have a solid understanding of what each of these words mean.

Reliability refers to the likelihood of getting the same results if you did the investigation again and being sure that the results are not just down to chance. Reliability is now often called repeatability for this reason. If you can repeat an investigation several times and get the same result each time, it is said to be reliable.

The reliability can be improved by:
- Controlling other variables well so they do not affect the results
- Repeating the experiment until no anomalous result are achieved
- Increasing precision

Precision indicates the spread of results from the mean.

The precision can be improved by:
- Using apparatus that has smaller scale divisions

Accuracy is a measure of how close the measured value is to the true value. The accuracy of the results depends on the measuring apparatus used and the skill of the person taking the measurements.

The accuracy can be improved by:
- Improving the design of an investigation to reduce errors
- Using more precise apparatus
- Repeating the measurement and calculating the mean

You can observe how these terms are used in the figure below.

Reliability v Precision v Accuracy

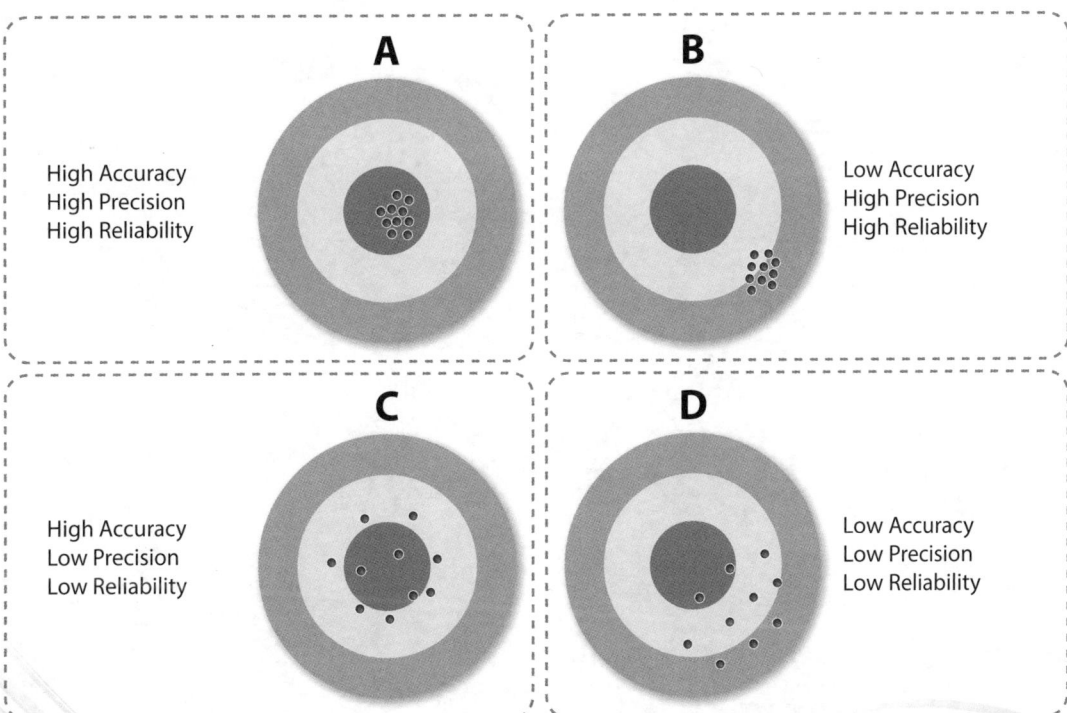

Quick skills section xiii

Designing an investigation

When asked to design an investigation you must think carefully about what level of detail to include. You must identify what your independent variable is and which values you are planning to use for it. The dependent variable must also be identified along with how you are going to measure it. The next thing you need to do is to suggest how you will control other variables. Finally, outline the method in a series of numbered steps that is detailed enough for someone else to follow. Remember to include repeat readings to help improve reliability.

1 Planet Earth

In this chapter, you will complete investigations on:

- ◆ 1.1   Estimating the percentage of oxygen in the air
- ◆ 1.2   The effects of acid rain on metal
- ◆ 1.3   The effect of carbon dioxide on the atmosphere

Practical investigation 1.1 Estimating the percentage of oxygen in the air

Objective

Oxygen is one of the gases that make up Earth's atmosphere. It is an important element and is involved in many reactions necessary for life. In this investigation you will estimate what percentage of the atmosphere is made up of oxygen. To do this you will use the rusting of iron. This is an oxidation reaction in which the oxygen from the atmosphere reacts with iron to form rust. As the oxygen is removed from the air in the tube to form iron oxide, the same volume of water is drawn up the tube. By measuring how far water is drawn up inside a boiling tube it will be possible to calculate the percentage of oxygen in the air. By the end of this investigation you should be able to state the composition of clean, dry air.

Equipment

- Iron wool
- boiling tube
- rubber bung
- beaker (250 cm^3)
- measuring cylinder (25 cm^3)
- glass rod
- permanent marker pen

Method

1. Place a piece of iron wool into the bottom of the boiling tube. It needs to be large enough so that is does not move when the boiling tube is inverted (turned upside down). You can use the glass rod to help you push the iron wool to the bottom.
2. Pour water into the boiling tube until it is half full so that the iron wool is completely covered then place the rubber bung on the boiling tube.
3. Fill the beaker with approximately 150 cm³ of water.
4. Invert the boiling tube and place it into the beaker, holding the bung while you do this to avoid the bung falling out.
5. Carefully remove the bung from the boiling tube (Figure 1.1).

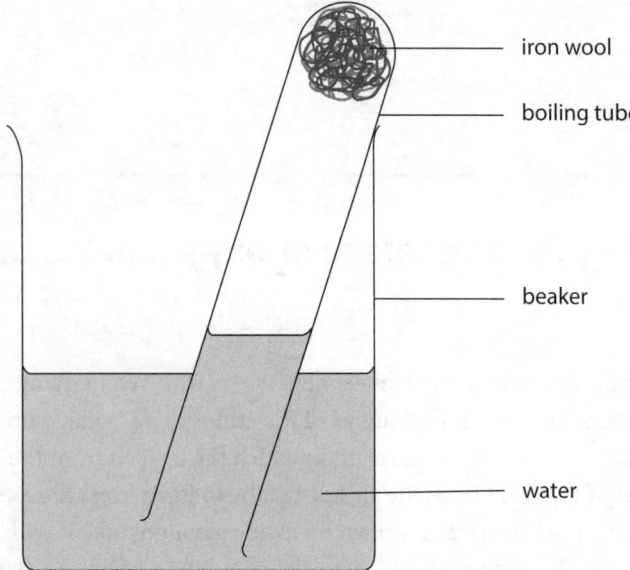

Figure 1.1

6. Using the marker pen, draw a line to mark how far the column of air is on the boiling tube. Label this line with the word 'start'.
7. Leave the experiment for a week.
8. Using the marker pen, a draw a line to mark the new position of the column of air. Label this line with the word 'finish'.
9. Carefully remove the boiling tube from the beaker.
10. Fill the boiling tube with water up to the line you labelled 'start'. Now pour the water into a measuring cylinder. Record this value in the table below.
11. Empty the measuring cylinder then fill the boiling tube with water to the line labelled 'finish'.
12. Now pour the water into a measuring cylinder. Record this value in the table. Also record any observations.

Safety considerations

None

Recording data

Because the water will have the same volume as the air inside the tube, by measuring the volume of water you will also be measuring the volume of air that was inside the tube.

Volume to start line / cm³	Volume to finish line / cm³
J 5.25 ml	4.1 ml
O 4.5 ml	3.75 ml

Observations of the iron wool inside the tube:

....

....

Handling data

You now need to calculate the difference between the two volumes. To do this you must subtract the volume of water to the finish line from the volume of water to the start line.

Difference in volume = ...

Now you need to calculate the % difference. To do this you must first divide the difference in volume that you calculated above by the value for the volume of water to the start line.

Difference in volume / Volume at start = ...

Now multiply this value by 100 to get the percentage = ...%

Analysis

1. Use the words below to fill in the gaps in the paragraph, and your own results to complete the conclusion.

   difference   iron oxide   rusting   water   iron   volume   oxygen

   The reaction that took place was called .... This is the name for the oxidation of .... Oxygen from the atmosphere reacted with the iron wool to form a new compound called .... or rust. At the start of the investigation the air in the tube contained .... but by the end this had all gone as it had reacted with the iron. As the oxygen had been removed from the air the .... of gas inside the tube was lower. This meant that .... from the beaker moved up inside the boiling tube. By measuring the .... in the volume of gas from start to finish, it was possible to calculate the volume of oxygen that had been removed from the air in the tube. This was then calculated to be ....%.

1 Planet Earth   3

Evaluation

2 Apart from the decrease in the volume of gas inside the tube, what other evidence was there for a reaction having taken place?

....

....

3 Why did the experiment have to be left for a week?

....

....

4 What other apparatus could you have used to measure more precisely the volume of oxygen being used up?

....

....

5 Use your textbook or the internet to find out the real percentage of oxygen in the air. Look at your estimate. Suggest why your estimated value was higher or lower than the real percentage of oxygen in the air.

....

....

....

....

6 Use the knowledge you have gained from this investigation and your general chemistry knowledge to suggest a method for estimating the percentage of carbon dioxide in the air.

....

....

....

....

Practical investigation 1.2 The effects of acid rain on metal

Objective

Acid rain caused by burning fossil fuels which contain sulfur compounds is a problem that increases the corrosion of metals. In this investigation, you will determine which metals are most affected by acid rain. Sulfur dioxide released by burning impure fossil fuels dissolves in water in the Earth's atmosphere to form acid rain. You will be using a sulfur dioxide solution to create a similar acidic atmosphere inside a container. By the end of this investigation you should be able to state how sulfur dioxide, from the combustion of fossil fuels which contain sulfur compounds, leads to acid rain.

Equipment

- Zinc foil
- tin foil
- iron sheet
- magnesium ribbon
- copper foil
- foam sheet
- plastic box with lid, beaker (100 cm^3)
- scissors, measuring cylinder (50 cm^3)
- marker pen
- sandpaper

Method

1. Place the foam sheet in the plastic box. You might need to use the scissors to cut it to the correct shape. Use the marker pen to mark the lid with the name of each metal. Make sure the labels are spaced out.
2. Carefully clean each piece of metal with the sandpaper and insert each piece into the foam so that it is standing upright. Ensure that each metal is next to the label you have written. Make sure each metal strip is secure and will not fall over.
3. Take the plastic box to the fume cupboard with the 100 cm^3 beaker.
4. Carefully measure 30 cm^3 of sulfur dioxide solution using the measuring cylinder and pour this into the 100 cm^3 beaker.
5. Place the beaker with the sulfur dioxide solution into the plastic box and seal the lid tightly.
6. Leave the plastic box inside the fume cupboard and return periodically to observe the effects of the sulfur dioxide on the metals.

Safety considerations

Wear eye protection throughout. Make sure that you do not remove the acid solution or the plastic boxes from the fume cupboard. Do not open the plastic box once you have placed the beaker of acid inside.

Recording data

In the space below you must design a results table that includes space for the names of each metal and space to record your observations of the effect of the sulfur dioxide solution. Ask your teacher how many times you will observe the results. If you need a reminder of how to draw a results table, have a look at the 'Quick skills guide' at the start of this book.

Analysis

1 From your observations, which of the metals were most corroded by the acid rain?

   ....

   ....

2 Which of the metals corroded first?

   ....

3 Why do you think that this metal corroded first?

   ....

   ....

4 Were there any metals that did not corrode?

   .... think ↓ ....

5 Why do you think that these metals did not corrode? (Alu, copper

   ....

   ....

6 List three ways that metals that corrode when exposed to acid rain could be protected?

   ....

   ....

   ....

7 Suggest two ways that a reactive metal used as a building material could be protected in an area that has a lot of acid rain.

   ....

   ....

   ....

Evaluation

8 Suggest why it was necessary to clean the metals with sandpaper at the start of the investigation.

   ....

   ....

9 Why was it possible to see corrosion on many of the metals in this investigation in only a few days when building materials can last for many years in areas where there is acid rain?

....
....
....

10 How could you have designed a control for this investigation?

....
....
....

Practical investigation 1.3 The effect of carbon dioxide on the atmosphere

Objective

Some gases in the atmosphere cause energy from the Sun to be trapped in the Earth's atmosphere rather than being allowed to pass out into space. If the level of these gases in the atmosphere increases, the temperature of the Earth will increase. This is called global warming. In this investigation, you will see if changing the level of carbon dioxide has an effect on the amount of heat energy stored in the gases inside a bottle. By the end of this investigation you should be able to state that carbon dioxide is a greenhouse gas and explain how it may contribute to climate change.

Equipment

- Two large clear plastic drinks bottles (at least 1.5 dm^3)
- two thermometers, two rubber bungs with holes for thermometers (or modelling clay)
- lamp, meter rule, measuring cylinder (1000 cm^3 or 500 cm^3)
- two antacid tablets, marker pen

Method

1. Using the measuring cylinder, measure 750 cm^3 of water into each of the bottles. Label one 'Carbon dioxide' and the other 'Normal air'.
2. Insert the thermometers into the bungs or use the modelling clay to create a lid for each bottle with the thermometer suspended inside.
3. Measure the temperature of the air in both bottles.
4. Add the antacid tablets to the water in the 'Carbon dioxide' bottle and close the bottle immediately with the bung or modelling clay. It is important that both bottles are sealed well so that they are airtight.
5. Place the lamp 40 cm from the bottles and switch it on.
6. After 45 minutes, record the temperature in each bottle again.

Safety considerations

The lamp may become hot as it will be on for a long time. Take care with thermometers as they can break very easily.

Recording data

Bottle	Temperature at the start/°C	Temperature at the end/°C
carbon dioxide		
normal air		

Handling data

Use the data from your results to plot a bar chart.

Analysis

1   For each of the two bottles, calculate the temperature change from the start to the end of the experiment.

Bottle	Temperature change from start to end / °C
carbon dioxide	
normal air	

2   Look at your results. What can you conclude from your investigation about the effect of carbon dioxide on the temperature of the gas inside the bottle?

....

....

....

Cambridge IGCSE Chemistry

Evaluation

3 List at least three variables in this experiment that you controlled.

   ....

   ....

4 How could you test the gas produced by antacid tablets to make sure it is carbon dioxide?

   ....

   ....

5 Why is it important that both bottles are the same distance from the lamp?

   ....

   ....

6 What would be the effect on the temperature of a bottle that was placed closer to the lamp?

   ....

   ....

7 How could you redesign this experiment to determine whether increasing carbon dioxide concentration increased the change in temperature?

   ....

   ....

   ....

   ....

Exam-style questions

1 Ravi and Jose are investigating the percentage of oxygen found in the air. They have left some wet iron wool in an inverted boiling tube like the one shown in Figure 1.2. Ravi is not sure that the reaction is complete; he thinks that there is still oxygen in the tube.

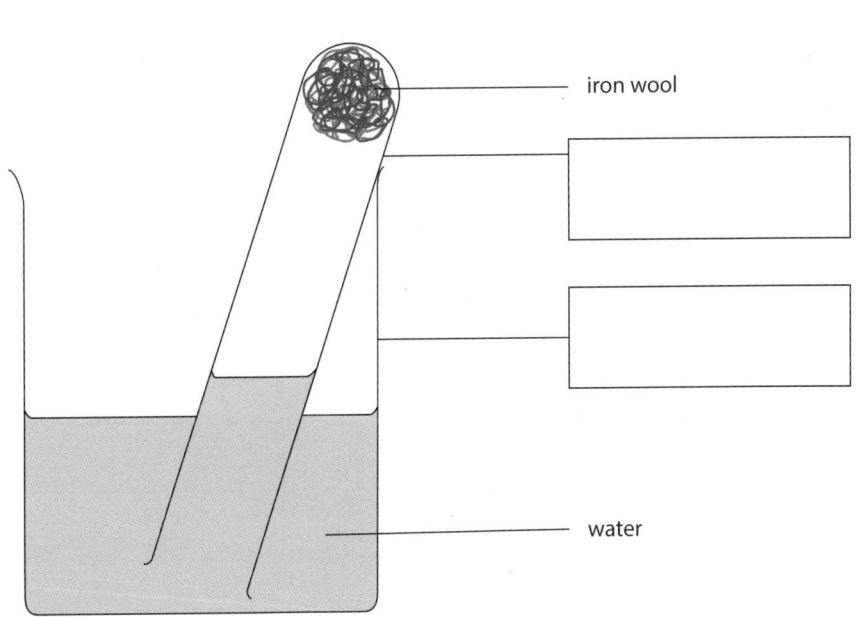

Figure 1.2

1 Planet Earth   11

a What can he do to make sure the reaction is over? [1]

....

....

b Complete the boxes to name the apparatus shown in the diagram. [2]

To get more accurate data, Ravi and Joe repeated the experiment using a different set of apparatus. This time they placed the wet iron wool into a conical flask and attached it to a gas syringe open to 100 cm³ to measure the decrease in volume.

c Use the gas syringe diagrams below to complete the table of results and calculate the decrease in the volume of gas. [8]

Repeat	Gas syringe diagram	Reading / cm³	Decrease in volume / cm³
1	30 40 50 60 70 80 90 100		
2	30 40 50 60 70 80 90 100		
3	30 40 50 60 70 80 90 100		
4	30 40 50 60 70 80 90 100		

d Calculate the mean decrease in the volume of gas. [2]

....

e As a control experiment, Ravi and Joe attached the gas syringe to a conical flask with dry iron wool and left it for three days. What do you think the decrease in volume of gas was? [2]

....

Total [15]

2 Mark and Tony are discussing the results of their investigation into the effect of acid rain on different metals. In their investigation, they used different metals and observed the effect of corrosion. Look at their results below.

Name of metal	Level of corrosion after 14 days
silver	none
lead	some corrosion
aluminium	very corroded

From the results, Mark concludes that silver should be used as a building material in areas where there is a lot of acid rain. Tony disagrees and says he can think of two reasons why using silver as a building material is a bad idea.

a Suggest two reasons Tony could use to support his argument. [2]

....

....

....

b From the results, suggest which of the metals is most reactive. [1]

....

c Predict how much corrosion would be observed if the experiment was repeated using magnesium. [1]

....

d Suggest how you could measure the pH of the acid rain that was used. [1]

....

....

Many lakes in Norway are affected by acid rain and so the government is monitoring the pH of many lakes. Bjorg and Peter are sent a sample of water from a nearby lake to test using a pH meter. They recorded the pH of the samples sent every day for 14 days in a table:

Day	pH of sample
1	6.5
2	6.4
3	6.5
4	6.6
5	5.5
6	5.5
7	5.6
8	5.7
9	5.8
10	6.9
11	7.0
12	7.0
13	6.9
14	6.9

e Plot the results for the experiment on the grid. [5]

f Using the results and your graph, suggest which day it rained. [1]

....

....

As part of the work to prevent damage caused by acid rain, the government of Norway is adding lime to the lake to reduce the acidity.

g Using the results and your graph, suggest which day lime was added to the lake and give a reason for your answer. [2]

Total [13]

3 Metals are a commonly used building material and so are very vulnerable to the damage that is caused by acid rain. However, it is possible to protect metals from this corrosion. Design an investigation to safely determine the best method of protecting metal from acid rain. Use the equipment list below. Remember to think about variables that you will control (keep the same).

Equipment:
- metal strips
- grease
- paint
- paintbrush
- cling film
- sulfuric acid solution
- foam, plastic box [5]

Total [5]

2 The nature of matter

In this chapter, you will complete investigations on:

◆ 2.1   Changing physical state

◆ 2.2   Filtration, distillation and evaporation

◆ 2.3   Chromatography

Practical investigation 2.1 Changing physical state

Objective

In chemistry, there are three states of matter: solid, liquid and gas. By changing the temperature of an element it is possible for us to change the state that it exists in. For example, if we heat water to 100 °C, it will begin to boil and change into a gas. Likewise, if we cool water to 0 °C, it will freeze and turn into ice. When changing states, energy is required to break the intermolecular forces between molecules. In this experiment, we will examine what happens to the temperature of water as it is heated from ice until it becomes steam. By the end of this investigation you should be able to describe changes of state in terms of melting, boiling, evaporation, freezing and condensation.

Equipment

- Clamp stand with clamp and boss
- heat-resistant mat
- Bunsen burner, thermometer
- beaker (250 cm³)
- ice
- timer
- pestle
- mortar
- tripod
- gauze
- stirring rod

Method

1. Add seven ice cubes to the mortar and crush them with the pestle until you are left with only small pieces. Do this carefully so that the ice cubes do not come out of the mortar.
2. Place the crushed ice in the beaker until it is half full.
3. Set up the Bunsen burner on the heat-resistant mat.
4. Place the beaker on the tripod and gauze. Use the clamp and clamp stand to hold the thermometer in the beaker. You can use the diagram in Figure 2.1 to help you.
5. Measure the temperature of the ice in the beaker and record the result in the results table.
6. Start the timer. Begin to heat the beaker of water with the Bunsen burner on a gentle blue flame.
7. Record the temperature every minute. Use the stirring rod to make sure the ice melts evenly.
8. Once the water is boiling (bubbles forming within the liquid), only take one more reading.

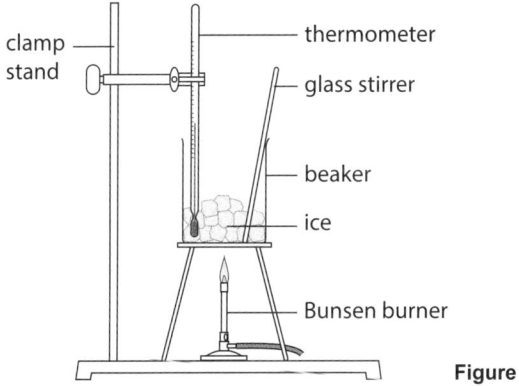

Figure 2.1

Safety considerations

Wear eye protection throughout. You will need to stand for the practical because hot liquids will be used. Remember to take care when handling hot glassware and also to be careful when the water is boiling as the steam will be very hot.

Recording data

1. Record your results in the results table below. The units are missing and need to be added.

Time /	Temperature /
0	
1	
2	
3	
4	
5	
6	
7	
8	

Handling data

2 Plot the results of your experiment on the graph paper below. Think about whether you will need to plot a line graph or a bar graph.

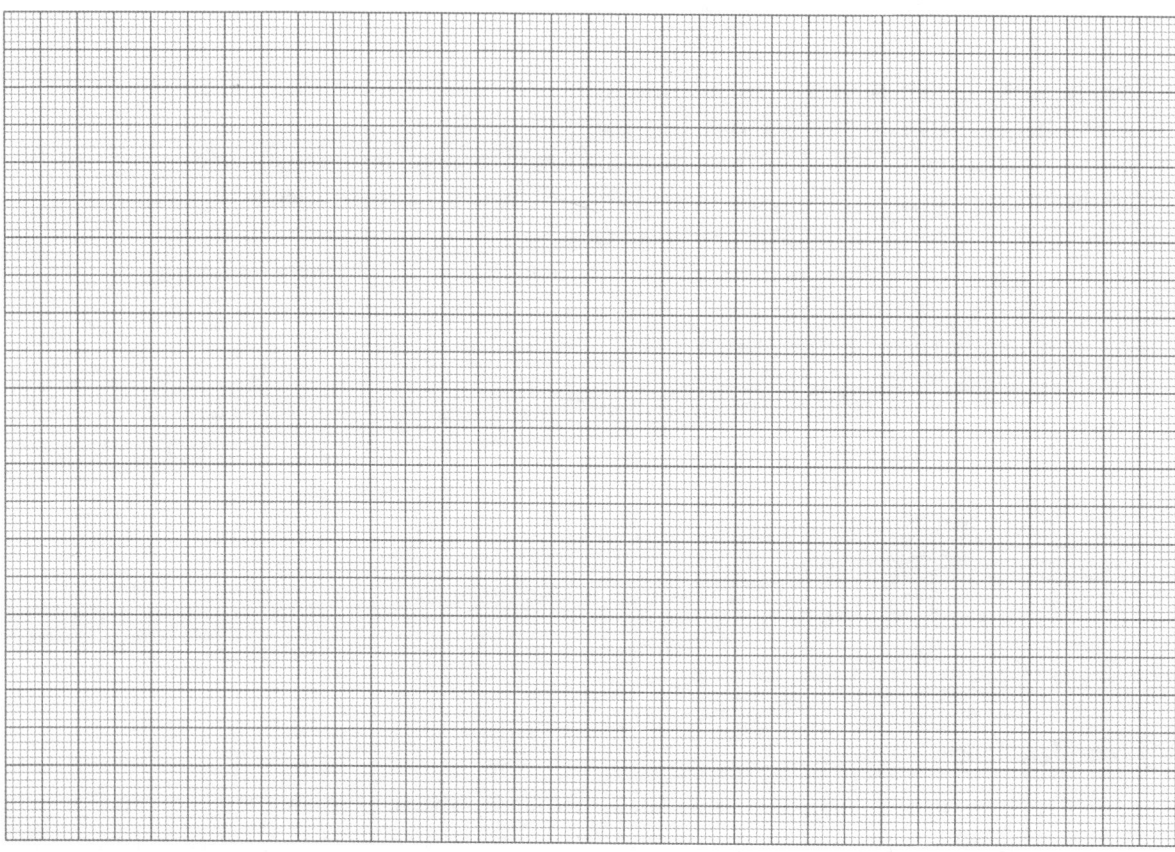

Analysis

3 Use the words given to complete the conclusion paragraph below.

> molecules   intermolecular   melting   temperature   boiling   liquid   gas   heating

At first, the .... inside the beaker did not change. This is because the energy being added by .... was being used to break the intermolecular forces between the water .... This is called .... Once all of the solid water had turned into .... water, the temperature began to increase. It stopped increasing once the water reached its .... point. The energy being added was now used to break the .... forces between the water molecules in the liquid state. This meant that the water could turn into a ....

18 Cambridge IGCSE Chemistry

4 Look at the graph in Figure 2.2.

Match the correct letter to each of the following words or terms:
- **a** melting point ....
- **b** boiling point ....
- **c** solid ....
- **d** liquid ....
- **e** gas ....

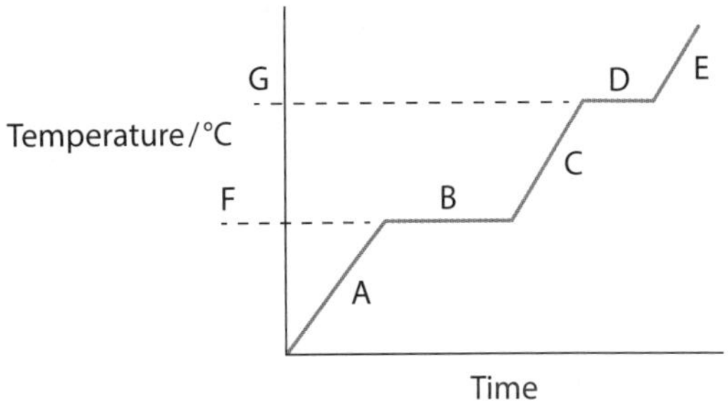

Figure 2.2

Evaluation

5 Think about your experiment. How could you have made the results more accurate?

....
....
....
....

6 How would adding an impurity, such as salt, to the ice in this experiment affect the results? Sketch a graph to support your ideas.

....
....
....
....
....

2 The nature of matter

Practical investigation 2.2 Filtration, distillation and evaporation

Objective

Often in chemistry we find that substances we want to use are in mixtures rather than in a pure state. It is important that we are able to separate these mixtures regardless of whether the substances mixed are solids or liquids. At a very basic level we might need to separate water from a salt solution to obtain drinking water, for example. At a more complex level we might want to separate the different hydrocarbons found in crude oil. In this investigation, you will begin with a mixture of four substances that have been mixed together. You will need to use a variety of different separating techniques to obtain samples of each substance. Your teacher will supply you with a sample of water that will have iron filings, sand and salt in it. You will need to think carefully about which techniques you will use to separate each substance from the others and also which order to complete the separations, as this is very important. By the end of this investigation you should be able to suggest suitable purification techniques, given information about the substances involved.

Equipment

- Filter paper
- clamp stand with clamp and boss
- funnel
- spatula
- two beakers (250 cm^3)
- magnet
- small plastic bag

- boiling tube with bung and delivery tube or Liebig condenser (if available)
- ice
- evaporating basin
- sample to be separated (of sand, salt, iron filings and 150 cm^3 of water)

Method

1 For each of the combinations below, suggest the most suitable technique for separating the substances.

   a To obtain salt from a salt water solution

   ....

   b To obtain iron filings from sand

   ....

   c To obtain water from a salt water solution

   ....

   d To obtain iron filings and sand from a water suspension

   ....

2 Now think about the order in which you will need to use each technique. Write the sequence of techniques in order.
   a ....
   b ....
   c ....
   d ....

3 You need to be able to describe methods of purification as part of the syllabus. For each of the techniques you have listed above, write a short method to explain how you will use the equipment to separate the mixtures.

Separating insoluble solids from liquids by filtration
a ....
b ....
c ....
d ....
e ....

4 For this section you will need to split the filtrate you have obtained into two samples: one for use in each of the two remaining techniques. Describe the methods you would use.

Separating water from a solution that contains a dissolved solid
a i ....
ii ....
iii ....
iv ....
v ....

Separating a soluble solid from water to obtain the solid
b i ....
ii ....
iii ....
iv ....
v ....

Separating two solids from one another
c i ....
ii ....
iii ....

Safety considerations

Wear eye protection throughout. You will need to stand for the practical because hot liquids are being used. Remember to take care when handling hot glassware and also to be careful when the water is boiling as the steam will be very hot.

Recording data

Standard diagrams are used to show the apparatus needed in chemistry experiments (see Figure 2.3 on the next page).

2 The nature of matter

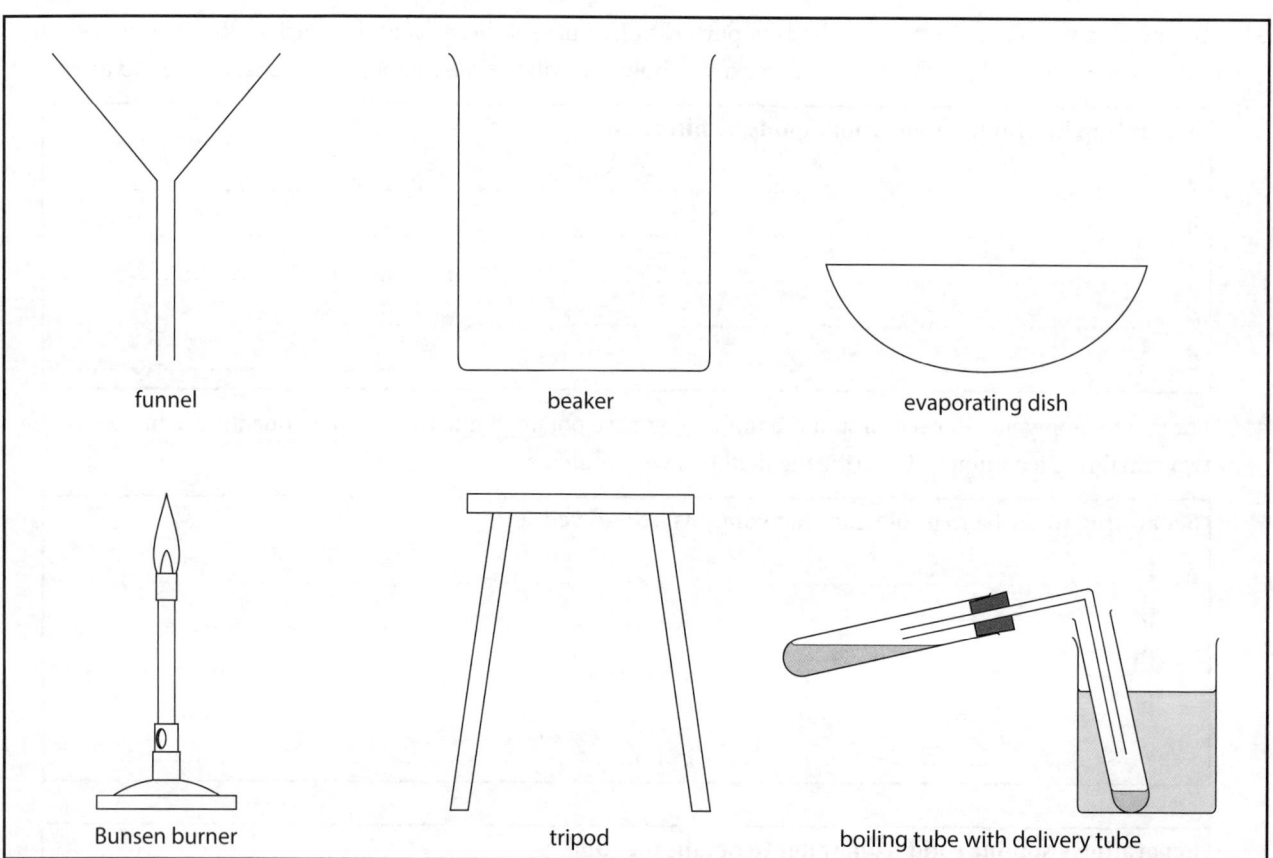

Figure 2.3

5 Refer to the illustration above and, in the spaces below on the opposite page, draw diagrams to show how you arranged your practical equipment.
   a Filtration diagram

22 Cambridge IGCSE Chemistry

b Evaporation diagram

c Distillation diagram

Evaluation

6 How could we test the water to make sure that there was no salt in it? (Important: Remember that you cannot taste anything in a laboratory.)

....

....

Practical investigation 2.3 Chromatography

Objective

Sometimes more than one solid is dissolved in a solution and so it is difficult to separate these from one another, therefore we use a technique called chromatography. This technique is used extensively in forensic science and sports for drug testing. By the end of this investigation you should be able to demonstrate knowledge and understanding of paper chromatography so that you can interpret simple chromatograms including the use of R_f values.

Equipment

- Beaker (250 cm^3)
- chromatography paper/filter paper
- samples of food dye
- paper clip
- capillary tube

Method

1. Read the **incorrect** method that Atikah has written in the box below. There are six errors. Underline these errors and then rewrite the method in the space provided with your corrections.

Judy's method

1. Take the filter paper and, using a ruler, measure 2 cm from the bottom. Use a pen to draw a line across the paper from one side to the other.

2. At 0.5 cm intervals label the filter paper with the full name of each sample of food dye that will be used.

3. Add a large sample of each of the types of food dye to the filter paper using a pipette.

4. Bend the filter paper into a circle and fix it in place with a paper clip.

5. Place the filter paper in the empty beaker so that it is touching the sides of the beaker.

6. Add water to the beaker until it is just above the line that you have drawn on the filter paper.

7. When the ink has climbed about three-quarters of the way up the paper, remove the paper and dry it.

Cambridge IGCSE Chemistry

Corrected version of method

1 ....
....

2 ....
....

3 ....
....

4 ....
....

5 ....
....

6 ....
....

7 ....
....

Safety considerations

Food dye can stain the skin and clothing.

Recording data

For each sample of food dye you will need to measure the distance travelled by each of the spots. To do this you will need to use your line as a starting point and then measure the distance to the middle of each spot and also the total distance moved by the solvent (Figure 2.4). You may not need all of the rows in each table as this will depend on the number of different solids dissolved in each sample.

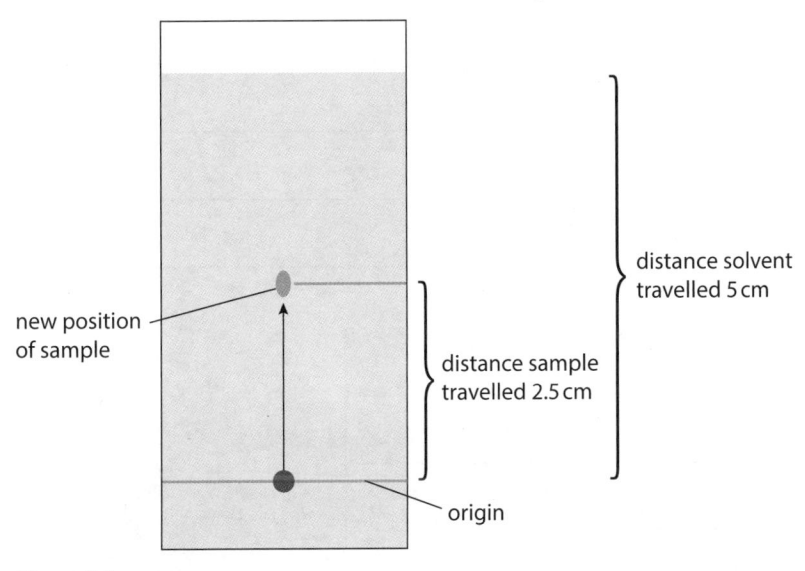

Figure 2.4

Total distance moved by solvent .... mm. This is known as the solvent front.

2 The nature of matter   25

Spot number	Distance moved by spot/mm

Sample A

Spot number	Distance moved by spot/mm

Sample B

Spot number	Distance moved by spot/mm

Sample C

Handling data

To allow comparison of different spots, we need to calculate the R_f value for each of the spots. This is very simple to do as all that is needed is a simple mathematical division. The distance travelled by each spot is divided by the distance travelled by the solvent. Complete the tables below.

Spot number	R_f value

Sample A

Spot number	R_f value

Sample B

Spot number	R_f value

Sample C

Analysis

2 With reference to the R_f values, identify which samples and spot numbers were the same.

....

....

....

3 Look at your samples. Was there a dissolved solid that was found in only one sample? Describe this with reference to the R_f value.

....

....

....

Evaluation

4 Identify a source of error in the investigation and how you would change the experiment to solve this problem.

....

....

....

Exam-style questions

1 Sweets often come in bright colours so that they appear attractive. These colours are made using water-soluble food dyes, which can be separated using the apparatus shown.

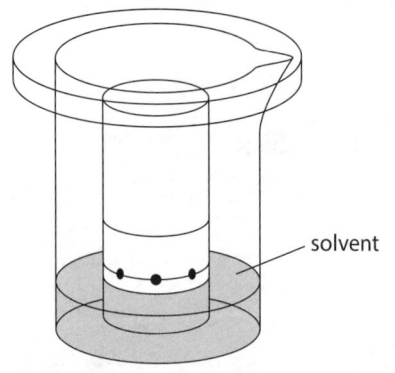

Figure 2.5

a What name is given to this separation technique? [1]

....

b Suggest a suitable solvent. [1]

....

c What would you use to draw the line at the bottom of the paper? [1]

....

d What is the name given to this line? [1]

....

Look at the diagram below. Three types of food dye were placed on the paper alongside the results for two types of sweets.

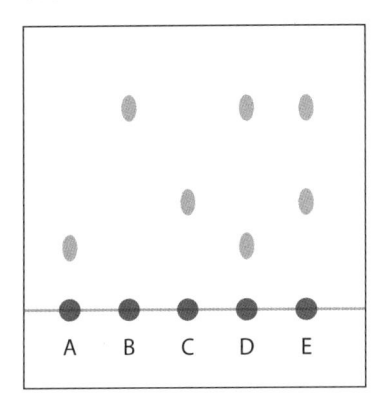

Figure 2.6

e i Which dyes are used to make sweet D? [2]

....

ii Which dyes are used to make sweet E? [2]

....

Total [8]

2 Some scientists were investigating the effect that a new type of antifreeze has on the boiling point of water.

a Use the thermometer diagrams in the table to complete the results column for the boiling point temperature. [6]

Mass of antifreeze added / g	Thermometer diagram	Boiling point / °C
0	105 / 100 / 95	
20	100 / 95 / 90	
40	95 / 90 / 85	
60	100 / 95 / 90	
80	85 / 80 / 75	
100	80 / 75 / 70	

b Plot the points on the grid and draw a smooth line graph. [6]

c Which of your results do you think is an anomaly? [1]

....

d Use your graph to find the boiling point of water with 90 g of antifreeze dissolved in it. [1]

....

e Suggest one control variable for this experiment. [1]

....

....

Total [15]

3 A mixture of two liquids with similar boiling points is separated using a method called fractional distillation. Look at the diagram, which shows how apparatus can be arranged to carry out this type of separation.

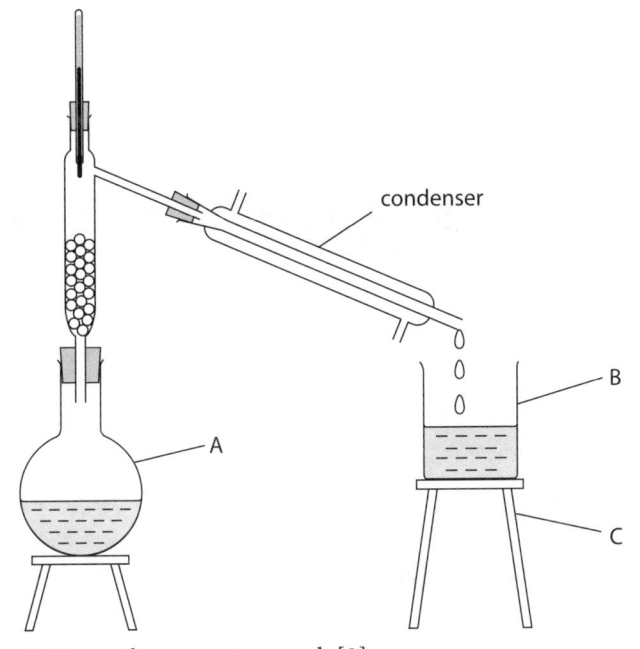

Figure 2.7

  a Name the apparatus used. [3]
  A ....
  B ....
  C ....

  b On the diagram, draw an arrow to show the direction that water would flow through the condenser. [1]

  c On the diagram, label where heat would be added to the apparatus in order to cause the mixture of liquids to boil. [1]

Total [5]

4 Alistair and Imogen are students who are investigating changes of state by heating ice cubes in a beaker using a Bunsen burner. They are recording the temperature every minute. They suggest four improvements to the method which they think will make their experiment more reliable.

  a Read their suggestions and put a tick in the box if you think that the suggestion will improve reliability. [2]

  Using a silver tripod instead of a black one ☐

  Repeating the experiment and calculating a mean temperature for melting and boiling point ☐

  Weighing the mass of ice used ☐

  Using a machine to stir the water instead of doing it by hand ☐

  b Find an example of where adding an impurity to water changes its melting or boiling point is used in everyday life. Explain why the melting point or boiling point is changed with a reference to the intermolecular forces between water molecules. [3]

  ....
  ....
  ....
  ....

Total [5]

2 The nature of matter

3 Elements and compounds

> **In this chapter, you will complete investigations on:**
>
> ◆ **3.1** The properties of metals and non-metals
>
> ◆ **3.2** The differences between elements and compounds
>
> ◆ **3.3** The properties of ionic and covalent compounds

Practical investigation 3.1 The properties of metals and non-metals

Objective

Elements in the Periodic Table are all classified as either metals or non-metals. Each group has its own properties that are used to classify an element. In this investigation, you will be given a number of unknown samples and will use a series of tests to determine whether a substance is a metal or a non-metal. By the end of this investigation you should be able to describe the differences between metals and non-metals.

Equipment

- Samples for testing
- pestle and mortar
- bulb
- battery pack/power pack
- three wires with crocodile clips

Method

1. First make a visual inspection of each sample and record your observations in a results table.
2. You need to set up a simple circuit to test whether each sample conducts electricity. In the space provided below, draw a diagram to show how you will arrange your circuit.

3. Place each sample into your circuit to determine whether it conducts electricity. Record your results in the table. Disconnect the circuit.
4. Touch each sample in turn with your fingertips. Does it feel cold or warm to the touch?
5. Place each sample in the pestle and grind it with the mortar. Do any parts break off?

Safety considerations

Make sure you wash your hands after touching all of the samples. Take care not to create a short circuit when testing the conductivity of the samples.

Recording data

You need to create a results table for this experiment. You need to think about how many different samples you have to test. The properties you will be testing are: appearance, conductivity of electricity, conductivity of heat, and malleability (whether the substance can change shape without breaking).

Analysis

1. From your results, write the names of the metals and non-metals in the table below.

Metals	Non-metals

2. For each substance, complete the conclusion sentences below:
   a I think that .... is a .... because

   ....

   ....

   b I think that .... is a .... because

   ....

   ....

   c I think that .... is a .... because

   ....

   ....

   d I think that .... is a .... because

   ....

   ....

   e I think that .... is a .... because

   ....

   ....

3 Use the words given below to fill in the blanks in the text.

> conductors   grey   brittle   yellow   malleability   metals   heat   non-metals   graphite

From my investigation I know that metals have an appearance that is .... in colour. Non-metals have a variety of colours including black and .... Metals are good .... of electricity but .... work well as insulators. An exception to this is ...., which a good conductor despite being a non-metal. Metals are also good conductors of .... as they feel cold to the touch while non-metals feel warm. .... was tested by hitting the substances with the pestle to see if any parts broke off. Non-metals are .... and so crumble easily. .... are malleable and can change shape in response to being hit.

Evaluation

4 Why is it difficult to draw conclusions about all metals and non-metals based on your investigation?

....
....
....

5 What other tests could you perform on the samples to compare the physical properties of metals and non-metals?

....
....
....

Practical investigation 3.2 The differences between elements and compounds

Objective

Elements are made of only one type of atom while compounds contain two or more different types of atom. The properties of elements can change once they become part of a compound. In this investigation, you will examine the properties of iron and sulfur and how these change once they are combined into a compound. By the end of this investigation you should be able to describe the differences between elements, mixtures and compounds.

Equipment

- Ignition tube/test-tube with iron powder and sulfur powder
- Bunsen burner
- tongs
- heat-resistant mat
- bar magnet
- sealed test-tube with a mixture of iron and sulfur

Method

1. Look at the mixture of iron and sulfur. Record your observations.
2. Using the bar magnet, test the mixture of iron and sulfur in the test-tube to see whether anything happens. You are looking to see if the mixture or either of its components is affected by the magnet, for example being pulled up the side of the test-tube.
3. Set up your Bunsen burner. Do not light it until your teacher tells you to do so.
4. Place the ignition tube into the tongs.
5. Heat the ignition tube on a blue flame until the mixture begins to glow. Once this happens, remove the ignition tube from the flame and place it on the heat-resistant mat. Be careful as the tube will be very hot.
6. Turn your Bunsen burner off.
7. Once your ignition tube has cooled, look at the appearance. Record your observations.
8. Use the bar magnet to test the new compound to see whether it is magnetic.

Safety considerations

The ignition tube will be prepared for you. Do not remove the mineral wool plug.

Recording data

Add the name of the compound formed to the equation and the table below and complete the table.

iron + sulfur → ....

Substance	Appearance	Magnetic or non-magnetic?
iron and sulfur mixture		

Analysis

1   What signs are there that a chemical reaction was taking place?

   ....

Evaluation

2   Why was there a mineral wool plug in the ignition tube?

   ....

   ....

3   Why might there still be some iron and sulfur inside the ignition tube?

   ....

   ....

4   Could this experiment be repeated replacing iron with copper? Give a reason to support your answer.

   ....

   ....

   ....

Practical investigation 3.3 The properties of ionic and covalent compounds

Objective

Ionic compounds usually form when a metallic element bonds to a non-metallic element. Covalent compounds usually form when two non-metallic elements bond. These two types of compound have very different properties. In this investigation, you will examine the physical properties of some ionic and covalent compounds. You will need to take careful observations of what you see. By the end of this investigation you should be able to describe differences in the melting point, solubility and electrical conductivity of ionic and covalent compounds.

Equipment

- Crucibles
- Bunsen burner
- heat-resistant mat
- clay triangle
- tripod
- spatula
- beaker (250 cm³)
- distilled water
- glass rod
- measuring cylinder (100 cm³)
- battery pack or 6 V power supply
- wires with crocodile clips
- 6 V bulb
- graphite rods
- two-hole rubber bung
- clamp and clamp stand
- emery paper
- samples of wax
- sugar
- sodium chloride *salt*
- silicon oxide *dioxide – silica*
- magnesium sulfate *epsom salt*
- zinc chloride

Method

You can complete each part of this investigation in any order.

Melting point

1. Place a small sample of the substance in a crucible. Place the crucible into the clay triangle. Heat the sample using the Bunsen burner on a blue flame. Stop as soon as the substance begins to melt. If the substance does not melt, try heating it with a roaring flame. If it still does not melt, record this in your results table. Be aware that some samples may 'spit' when heated.
2. Repeat step 1 with each of the other samples using a clean crucible each time.

Solubility

1. Add 100 cm³ of water to a beaker. Using a spatula, add a small amount of the substance. Stir with the glass rod. Record whether the substance dissolves.
2. Repeat using fresh water for each of the other samples.

Electrical conductivity in solution

1. Set up a simple circuit as shown in Figure 3.1.
2. Half fill a 100 cm³ beaker with water. Add a spatula of the substance. Insert the electrodes and then turn on the power supply. Record whether the bulb lights up.
3. Pour away the water and rinse out the beaker, then repeat for each of the other samples.

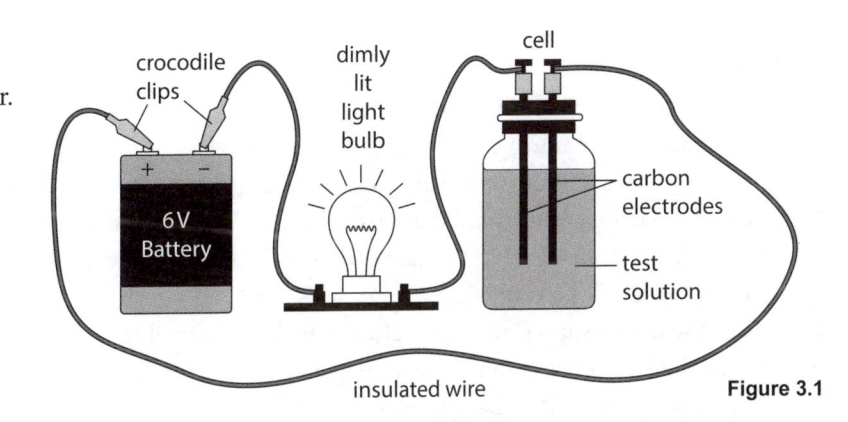

Figure 3.1

Electrical conductivity when solid

1. Set up a simple circuit as shown in Figure 3.2.

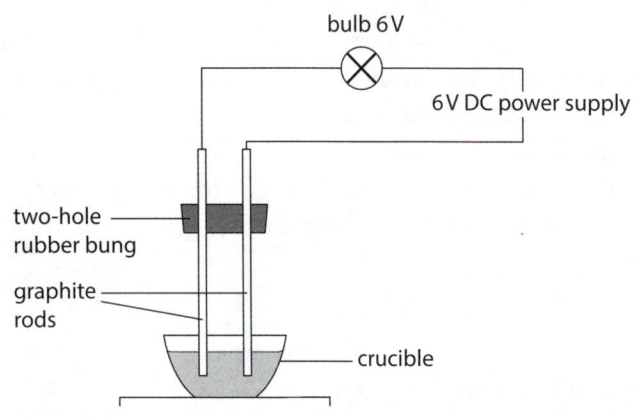

Figure 3.2

2. Add a few spatulas of the substance being tested to the crucible then test the substance by lowering the electrodes into the crucible and turning on the power supply. Record whether the bulb lights up, making sure the electrodes are not touching.
3. Turn the power supply off, remove the crucible and then repeat for each of the different samples.

Electrical conductivity when molten

1. Set up the simple circuit and apparatus shown in Figure 3.3.

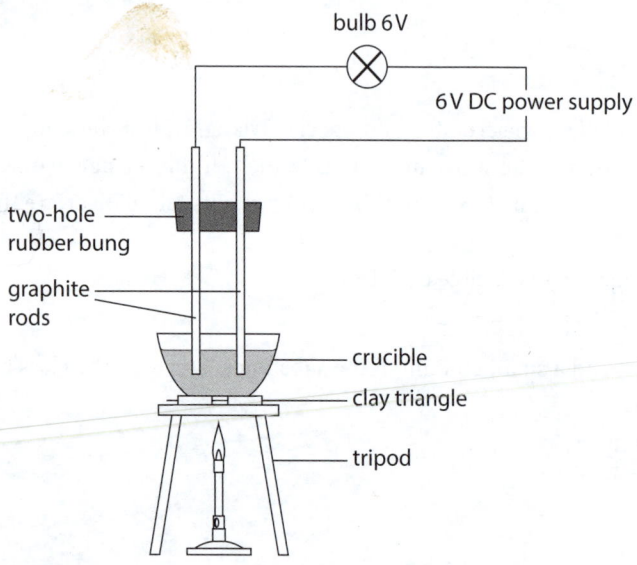

Figure 3.3

2. Half fill your crucible with sugar.
3. Place the crucible into a clay triangle on a tripod.
4. Prepare your electrodes and clamp stand so that it can be lowered into the crucible easily.
5. Heat the crucible gently until the substance just melts. Turn the Bunsen burner off and lower the electrodes into the molten substance. Turn on the power supply and record whether the bulb lights up. Take extra care as the molten solid will be very hot.
6. Turn the power supply off and carefully remove the crucible.
7. Clean the electrodes with the emery paper.
8. Repeat for each of the other samples.

Safety considerations

Wear eye protection throughout. Take care not to touch the graphite electrodes as you will create a short circuit. As soon as the zinc chloride melts, stop heating it otherwise chlorine gas may be produced.

Recording data

You will need to design a results table that will include all five substances being tested. There are also five different tests being done.

Handling data

1 For each of the substances tested, decide where they belong in the Venn diagram shown.

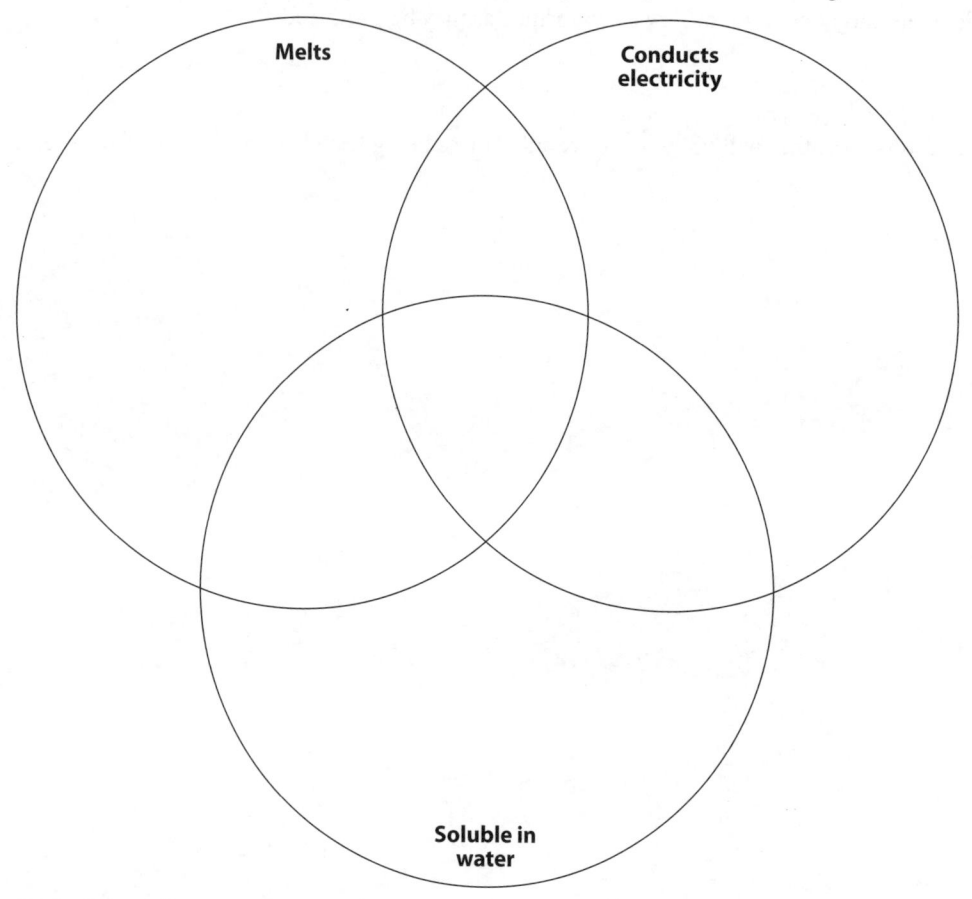

Melts

Conducts electricity

Soluble in water

2 One of the substances does not fit anywhere on the diagram. Which one is this?

....

Analysis

3 Fill in the table below.

The properties of ionic compounds	Substances tested that had these properties
* * *	* * *
The properties of covalent compounds	Substances tested that had these properties
* * *	* * *

42 Cambridge IGCSE Chemistry

Evaluation

4 The test for solubility was qualitative – it gave us a yes or no answer. How could you modify the test to get quantitative (numerical value) data?

....
....

5 The test for conductivity was qualitative – did the bulb light up or not? How could you modify the test to get quantitative (numerical value) data?

....
....

6 How might the temperature of the water you used have affected solubility?

....
....

7 One of the substances did not fit into the Venn diagram above. Decide which type of substance you think it is and give reasons for your choice.

....
....
....
....

Exam-style questions

1 One of the ways that materials can be tested to find out if they are metals or non-metals is to see if the material conducts electricity or not. Ola and Francesca set up a circuit like the one in Figure 3.4 and different materials were placed between the crocodile clips. The data is shown in the table.

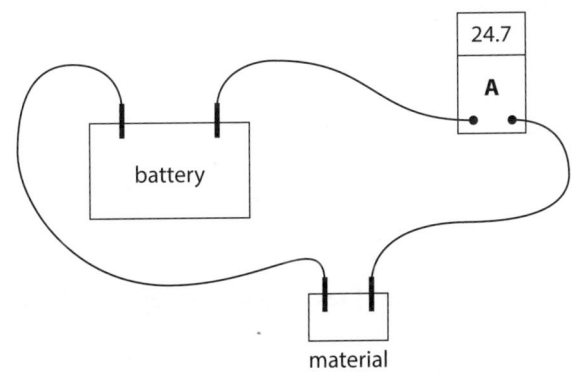

Figure 3.4

Material	Current/A
A	0.2
B	0.1
C	0.0
D	0.4
E	0.8

a Use the results to plot a bar graph on the graph paper below. [5]

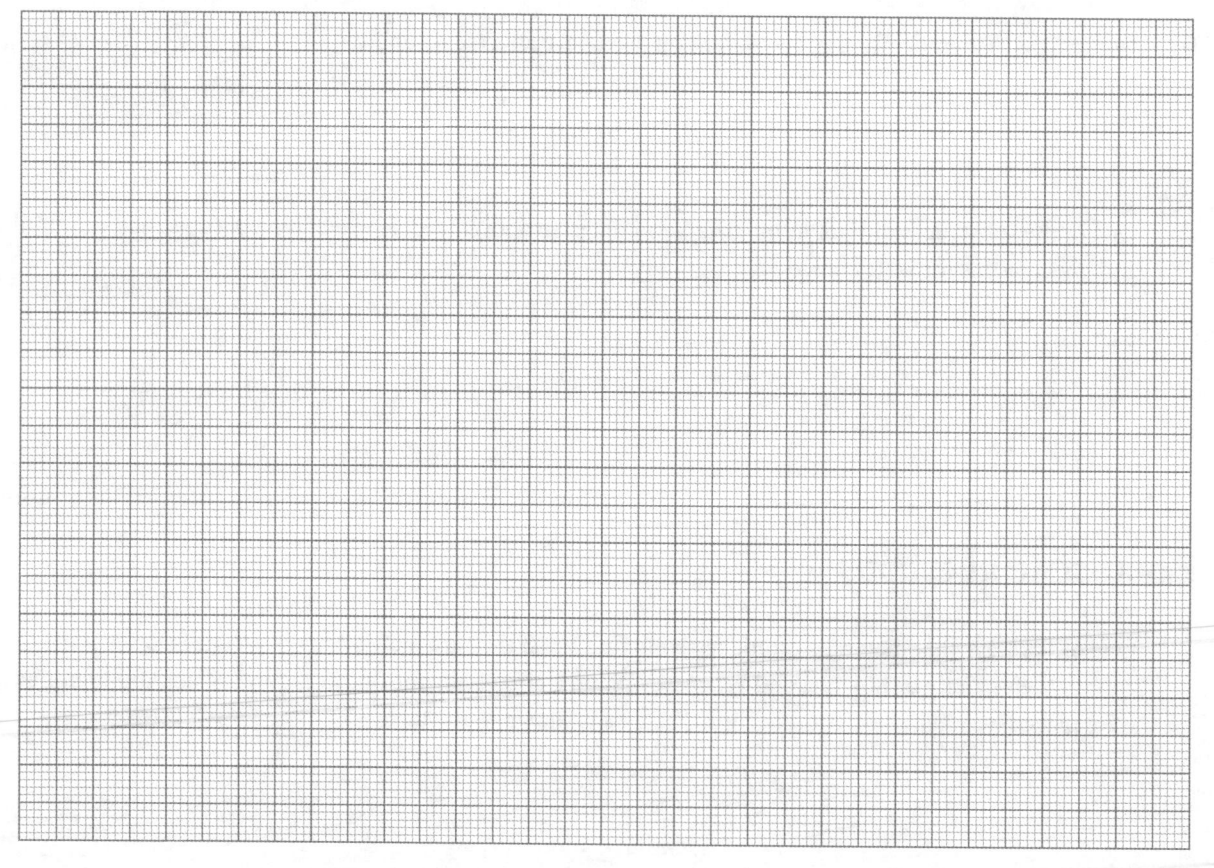

b One of the materials was a non-metal. Using the information in both the table of results and the graph you have drawn, suggest which letter it was. [1]

....

c One of the materials was gold. Using the information in both the table of results and the graph you have drawn, suggest which letter it was. [1]

....

44 Cambridge IGCSE Chemistry

d Suggest another circuit component that could have been used in this experiment instead of the ammeter. [1]

....

e Why is an ammeter a better component to use than your suggestion in **d**? [1]

....

Total [9]

2 Boris and Adina are trying to decide if a newly discovered compound is ionic or covalent. Plan an investigation to determine which type of compound it is. [6]

....

....

....

....

....

Total [5]

3 Magnesium is a very reactive metal which burns in air to form a white powder.
  a Look at Figure 3.5 and identify the apparatus used. [3]

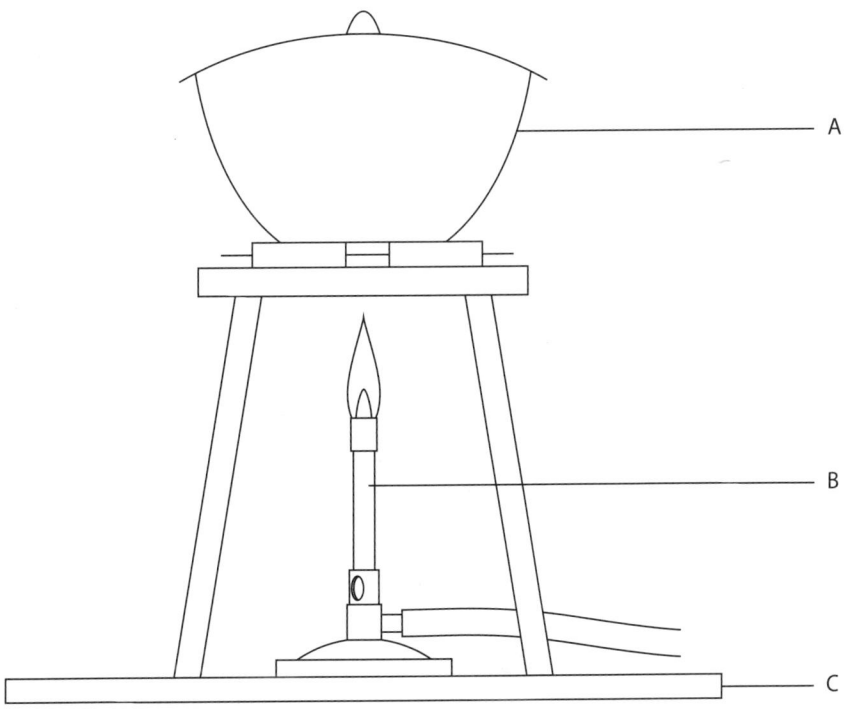

Figure 3.5

A ....
B ....
C ....

3 Elements and compounds 45

b On the diagram, draw an arrow to show where the magnesium should be placed. [1]

c What is the name of the gas that magnesium reacts with in the air? [1]

....

d Suggest one safety precaution that you would take during this experiment. [1]

....

....

e Describe how you could compare the mass of the contents of the crucible before and after the experiment. [3]

....

....

Total [9]

4 Chemical reactions

In this chapter, you will complete investigations on:

- 4.1 Types of chemical reaction
- 4.2 Exothermic and endothermic reactions
- 4.3 The electrolysis of copper

Practical investigation 4.1 Types of chemical reaction

Objective

Chemistry is often described as the study of change. It is important to understand that the changes that take place during chemical reactions are of a variety of types. In the investigation, you will examine a number of different types of chemical reactions. By the end of this investigation you should be able to identify physical and chemical changes, and understand the differences between them.

Equipment

Photosynthesis:	• Fresh leaves or pondweed (*Elodea*) • boiling tube, beaker (250 cm³) • lamp • rubber bung	• sodium hydrogen carbonate (bicarbonate of soda) • spatula	
Combustion:	• Magnesium ribbon • tongs	• Bunsen burner • heat-resistant mat	
Displacement:	• Beaker (100 cm³) • copper sulfate solution	• measuring cylinder (50 cm³)	• two zinc strips
Precipitation:	• Test-tube • sodium chloride solution	• silver nitrate solution • pipette	
Thermal decomposition:	• Test-tube • copper carbonate	• tongs • Bunsen burner	• spatula • heat-resistant mat

Method

There are five reactions that can be completed in any order.

Photosynthesis (Light)
1. Fill the beaker with water. Add 1 spatula of the sodium hydrogen carbonate and stir until it dissolves.
2. Place the leaf discs or pondweed (*Elodea*) into the boiling tube. Fill the boiling tube to the top and seal with a rubber bung.
3. Turn the boiling tube upside down and place it in the beaker. Remove the bung and allow the tube to rest in the beaker.
4. Place the lamp approximately 10 cm from the beaker and the turn it on. Leave for 5 minutes and then return to make observations.

Magnesium in air (Combustion)
1. Set up the Bunsen burner on the heat-resistant mat.
2. Hold the magnesium ribbon in the tongs.
3. Switch the Bunsen burner so that is on a blue flame and then hold the magnesium ribbon in the flame. Do not look directly at the magnesium when it burns as this is dangerous.
4. Record your observations.

Zinc in copper sulfate (Displacement)
1. Pour 50 cm^3 of copper sulfate solution into a beaker.
2. Place one of the zinc strips in the beaker and leave it for 5 minutes.
3. Compare the colour of the zinc that was submerged in the copper sulfate solution and the colour of the zinc that was not.
4. Record your observations.

Sodium chloride solution and silver nitrate (Precipitation)
1. Pour 15 cm^3 of sodium chloride solution into a test-tube.
2. Using a pipette, add 3 cm^3 of silver nitrate solution to the test-tube.
3. Record your observations.

Copper carbonate (Thermal decomposition)
1. Add a large spatula of copper carbonate to the test-tube.
2. Set up the Bunsen burner on the heat-resistant mat.
3. Using tongs, hold the test-tube in a blue flame and heat for 30 seconds.
4. Record your observations.

Safety considerations

Wear eye protection at all times and stand up while carrying out any reactions involving heating. Take care when heating the magnesium not to look directly at the flame. (Your teacher may supply you with a special filter to view the reaction through.)

Recording data

Reaction	Observations
leaf discs/pondweed (*Elodea*)	
burning magnesium	
zinc and copper sulfate	
sodium chloride and silver nitrate	
copper carbonate	

Analysis

1 For each of the reactions in the investigation where there were signs that a reaction was taking place, decide what evidence there was for a chemical reaction.

   a Leaf discs/pondweed (*Elodea*)

   ....
   ....

   b Burning magnesium

   ....
   ....

   c Zinc and copper sulfate

   ....
   ....

   d Sodium chloride and silver nitrate

   ....
   ....

   e Copper carbonate

   ....
   ....

Evaluation

2 Why was sodium hydrogen carbonate added to the solution before the leaf discs/pondweed were added?

....

....

3 Describe a control experiment for the pondweed practical that would show that light is needed for photosynthesis.

....

....

4 What would have happened if, instead of zinc, silver was added to the beaker containing copper sulfate solution?

....

....

5 Predict what gas was given off when copper carbonate was heated. Suggest a test for this gas.

....

....

6 In the space below, draw a mind map to show the signs you would look for to demonstrate that a chemical reaction had taken place.

Practical investigation 4.2 Exothermic and endothermic reactions

Objective

When chemical reactions take place, energy is either taken in or given out. Reactions where energy is required are called **endothermic**. Reactions that release energy into the surroundings are called **exothermic**. In this investigation, you will examine a number of reactions with the aim of determining which are endothermic and which are exothermic. To do this, you will measure the change in temperature from the start to the end of the reaction. By the end of this investigation you should be able to describe the meaning of exothermic and endothermic reactions.

Equipment

- Polystyrene cup and lid with hole
- beaker (250 cm^3)
- thermometer
- measuring cylinder (10 cm^3)
- spatula
- copper sulfate solution
- hydrochloric acid
- sodium hydrogen carbonate solution
- sodium hydroxide
- citric acid
- magnesium powder
- sodium carbonate
- ethanoic acid

Method

1. Place the polystyrene cup in the 250 cm^3 beaker. This will support it during the experiment.
2. Measure 10 cm^3 of sodium hydrogen carbonate solution and add it to the polystyrene cup. Place the lid on the cup and insert the thermometer through the hole (Figure 4.1). Use the thermometer to measure the temperature and record this in your results table.

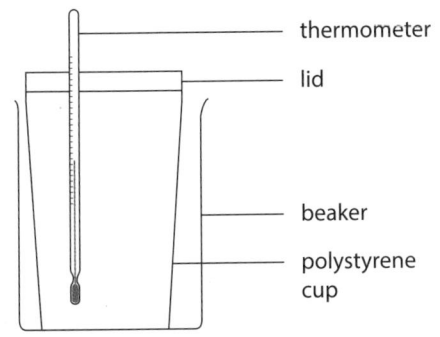

Figure 4.1

3. Add four spatulas of citric acid to the cup and cover with the lid immediately. Use the thermometer to stir the mixture. Record the maximum or minimum temperature achieved.
4. Pour the mixture away and rinse out the cup. Also rinse off the thermometer.
5. Measure out 10 cm^3 of sodium hydroxide solution and add this to the cup. Record the temperature.
6. Add 10 cm^3 of hydrochloric acid to the cup and cover with the lid. Use the thermometer to stir the mixture. Record the maximum or minimum temperature achieved.
7. Repeat step **4**.
8. Measure out 10 cm^3 of copper sulfate solution and add it to the polystyrene cup. Use the thermometer to measure the temperature and record this in your results table.
9. Add a spatula of magnesium powder to the cup and cover with the lid. Use the thermometer to stir the mixture. Record the maximum or minimum temperature achieved.
10. Repeat step **4**.
11. Measure out 10 cm^3 of ethanoic acid and add it to the polystyrene cup. Use the thermometer to measure the temperature and record this in your results table.
12. Add a spatula of sodium carbonate to the cup and cover with the lid. Use the thermometer to stir the mixture. Record the maximum or minimum temperature achieved.

Safety considerations

Eye protection must be worn at all times. Sodium hydroxide and citric acid are irritants.

Recording data

In the space provided below, design a results table. Consider how many columns you will need as you will be recording the starting temperature and the maximum/minimum temperature.

Handling data

For each of the reactions, calculate the temperature change. Remember to include a plus or minus sign in front of the temperature to show whether there was an increase or decrease.

Reaction	Temperature change / °C
sodium hydrogen carbonate and citric acid	
sodium hydroxide and hydrochloric acid	
copper sulfate and magnesium powder	
ethanoic acid and sodium carbonate	

Analysis

1 Which reactions were exothermic and which were endothermic?

　a The exothermic reactions were:

　...

　...

　b The endothermic reactions were:

　...

　...

Evaluation

2 Why was a polystyrene cup used?

....

....

3 What effect would not using a lid have had on the results?

....

....

4 If they were left for a longer period of time, all of the solutions would return to the original temperature. Where does the heat energy go?

....

....

5 Why was it necessary to stir the solutions?

....

....

6 How could you improve this investigation to get more accurate results?

....

....

Practical investigation 4.3 The electrolysis of copper

Objective

Electrolysis is used to obtain products that can be very valuable. There are some substances that can be very difficult to obtain in any other way. By the end of this investigation you should be able to relate the products of electrolysis to the electrolyte and electrodes used, exemplified by aqueous copper(II) sulfate using copper electrodes.

Equipment

- Beaker (250 cm³)
- two copper electrodes
- clamp stand and clamp
- D.C. power supply (6 V)
- leads and crocodile clips
- copper sulfate solution
- marker pen

Method

1. Pour 150 cm³ of copper sulfate solution into the beaker.
2. Mark one copper electrode with the letter A and the other copper electrode with the letter B at one end. Weigh each electrode and record the masses in the results table below.
3. Use the clamp stand to secure the two copper electrodes so that they are partially submerged in the copper sulfate solution. Connect the crocodile clips and leads to the power supply but do not switch it on yet. Ensure that your copper electrodes are not touching (Figure 4.2).

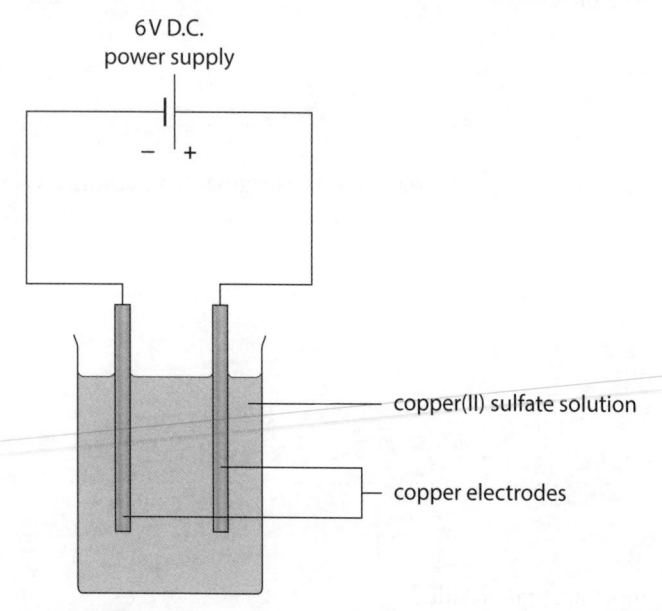

Figure 4.2

4. Make sure the power supply is set for 6 V and then switch it on.
5. Allow the reaction to run for 30 minutes.
6. Turn the power supply off. Carefully remove the copper electrodes from the solution. Allow them to dry for 5 minutes naturally. Do not use a paper towel to dry the electrodes.
7. Reweigh the copper electrodes. Record your data in the table.

Safety considerations

Wear eye protection. Ensure that the electrodes do not touch.

Recording data

Electrode	Mass at start/g	Mass at end/g	Change in mass/g
A			
B			

Handling data

1   What was the difference between the change in mass of electrode A and electrode B?

   ....

Analysis

2   From the data, which of the two electrodes was the anode and which was the cathode? Give reasons for your answers.

   ....

   ....

   ....

   ....

Evaluation

3   Why did you air-dry the copper electrodes rather than using paper towel to dry them?

   ....

4   What was the purpose of the copper sulfate solution?

   ....

   ....

5   Explain why the mass lost by the anode was greater that the mass gained by the cathode.

   ....

   ....

6   How would the results have been different if graphite electrodes had been used instead of copper ones?

   ....

   ....

4 Chemical reactions

Exam-style questions

1 Hessa and Kristen carried out an investigation to measure the temperature change when they mixed two solutions. This was their method:
- Add 20 cm³ of solution **A** to a plastic cup.
- Use a thermometer to record the temperature of solution **A**.
- Add 5 cm³ of solution **B** and record the highest temperature reached.
- Repeat the experiment using the same volume of solution **A** each time but increasing the volume of **B** used by 5 cm³ each time.

a Their results are given below in the table. Record the temperatures shown in the diagrams in the table. [7]

Volume of solution B added /	Thermometer diagram	Highest temperature reached /
5	(reads ~24)	
10	(reads ~27)	
15	(reads ~29)	
20	(reads ~32)	
25	(reads ~36)	
30	(reads ~35)	
35	(reads ~34)	

b The units are missing from the table. Add them in. [2]

c Suggest which volume of solution B could have been used as a control. [1]

....

d Why did the students use a plastic cup? [1]

....

e Plot the results on the graph paper below. [5]

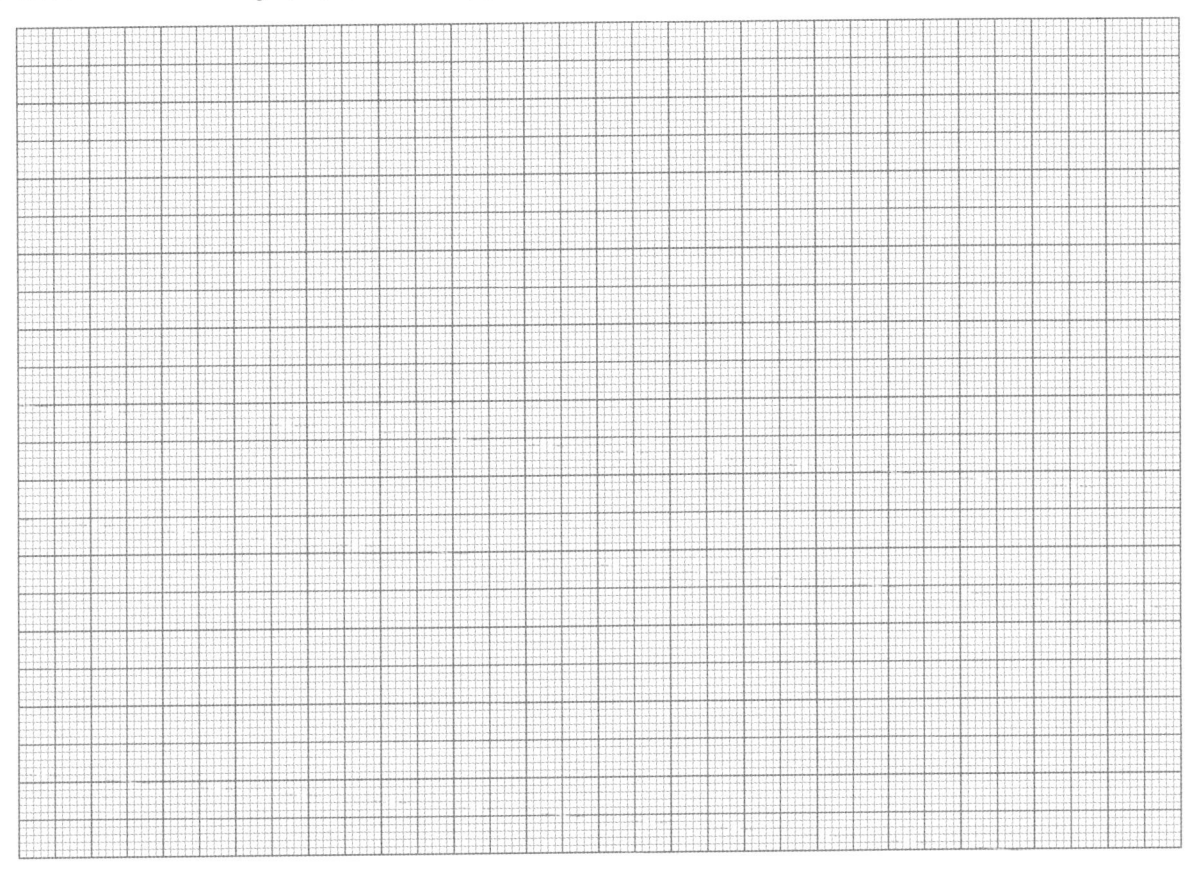

f The students each wrote a conclusion for the investigation. Read both of them and tick the one you think is correct. [1]

Conclusion 1: This was an exothermic reaction as the temperature increased when the two solutions were mixed. ☐

Conclusion 2: This was an endothermic reaction as the temperature decreased when the two solutions were mixed. ☐

g Suggest how this investigation could have been improved to prevent heat being lost to the surroundings. [1]

....

Total [18]

2 Copper can be extracted from its ore by reacting it with carbon. This is possible because carbon is more reactive than copper. Copper can also be extracted using the process of electrolysis.

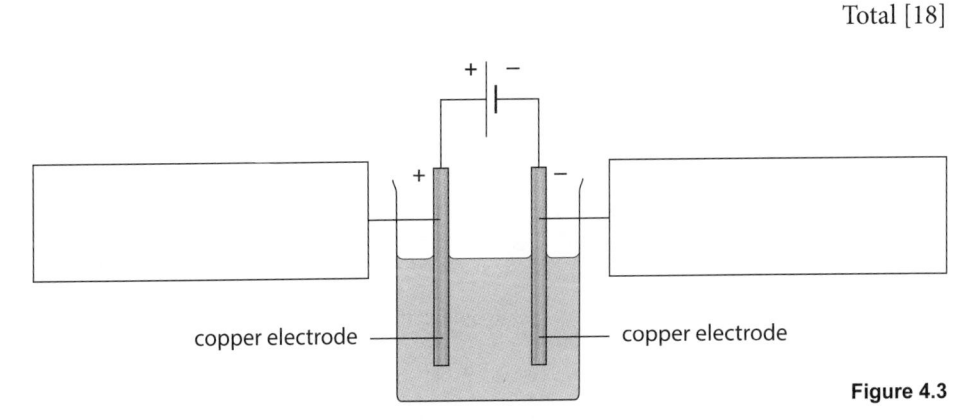

Figure 4.3

a Add labels to Figure 4.3 to show the anode and cathode. [2]

4 Chemical reactions 57

b Suggest a suitable solution that could be used. [1]

....

c What will happen to the mass of the cathode during this investigation? [1]

....

d What will happen to the mass of the anode during this investigation? [1]

....

e If copper is able to be extracted from its ore by using reduction, why is electrolysis also used in the production of copper? [1]

....

f No gas is produced during the electrolysis of copper. What could you add to the circuit to make sure that it was complete and that a reaction was taking place? [1]

....

g At the end of the investigation the students found brown sludge at the bottom of the beaker. What was this and where did it come from? [2]

....

Total [9]

3 Plaster of Paris is used to take imprints found at crime scenes. When water is added to the dry powder, a reaction takes place. Some students want to investigate whether this reaction is exothermic or endothermic. Describe an investigation that the students could do that would help them draw a conclusion. [6]

....

....

....

....

....

[Total 6]

4 When it is heated, copper carbonate decomposes into copper oxide and carbon dioxide.

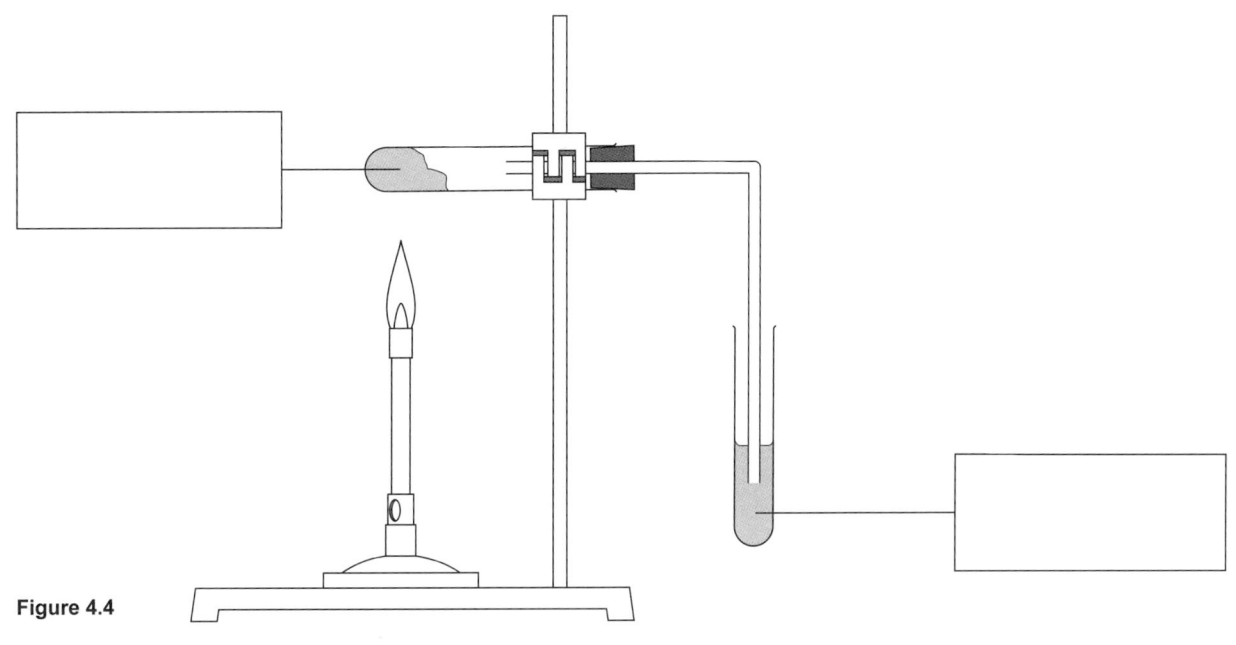

Figure 4.4

a Add a label to Figure 4.4 above to show where the copper carbonate is. [1]

Limewater can be used to test for the presence of carbon dioxide.

b Add a label to Figure 4.4 above to show where the limewater is. [1]
c Describe what you would see in the limewater if carbon dioxide is being produced. [1]

....

d What type of flame would you use to heat the copper carbonate? [1]

....

e How could you tell when the copper carbonate was completely decomposed to copper oxide? [1]

....

Total [5]

5 Kim and Stelios want to purify some copper using electrolysis.
   a Create a method for them to use. [6]

....

....

....

....

....

....

....

4 Chemical reactions 59

b Describe and explain why the mass of both electrodes changes during the reaction. [8]

....
....
....
....

Total [14]

5 Acids, bases and salts

In this chapter, you will complete investigations on:

- 5.1 Weak and strong acids
- 5.2 Reacting acids
- 5.3 Reacting alkalis
- 5.4 The preparation of soluble salts
- 5.5 The preparation of insoluble salts
- 5.6 Acid titration
- 5.7 The pH of oxides

Practical investigation 5.1 Weak and strong acids

Objective

Not all acids have the same strength – some are very weak and others are much stronger. Whether a substance is acidic or not depends on the ratio of hydrogen ions to hydroxide ions. If there are more hydrogen ions than hydroxide ions, the substance is acidic. In this investigation, you will compare two unknown acids and draw conclusions about their strength based on the results of the various tests. By the end of this investigation you should be able to describe the characteristic properties of acids as reactions with metals and bases and their effect on litmus and methyl orange.

Equipment

- 10 cm^3 of each of two acids
- sodium hydroxide solution
- sodium carbonate solution
- Universal Indicator solution
- Universal Indicator colour chart
- one piece of blue litmus paper
- methyl orange
- test-tubes
- test-tube rack
- four pipettes
- spatula
- two beakers (100 cm^3)
- matches
- splints
- dropping tile
- measuring cylinder (10 cm^3)
- two pieces of magnesium ribbon
- permanent marker

Method

1. Using a pipette, pour 2 cm³ of acid X into each of four test-tubes. Label these with an X and place them in the test-tube rack. Repeat for acid Y with four fresh test-tubes.
2. Measure 10 cm³ of sodium carbonate solution and pour it into a beaker. Do the same for the sodium hydroxide solution using another beaker.
3. Add two drops of Universal Indicator to two of the tubes containing acid X. Repeat for acid Y. Record the colour change in the table.
4. Using a clean pipette, slowly add drops of sodium carbonate solution to the first test-tube containing acid X and Universal Indicator. Count how many drops you have to add before the solution becomes neutral. You will need to shake the tube to get the solutions to mix properly. In the table below, record the number of drops needed and also a description of the reaction. Repeat for acid Y.
5. In the next two test-tubes containing acids X and Y with Universal Indicator, repeat the procedure but this time use a clean pipette and the sodium hydroxide solution. In the table, record the number of drops and also a description of the reaction. Remember to shake the tubes to ensure adequate mixing.
6. Take a piece of magnesium ribbon and add it to the next test-tube containing acid X. Record a description of the reaction in the table. Collect the gas given off using an empty test-tube upside down over the reaction tube. After about 30 seconds, place a lighted splint over the mouth of the upside down test-tube (Figure 5.1). Record the result. Repeat for acid Y.

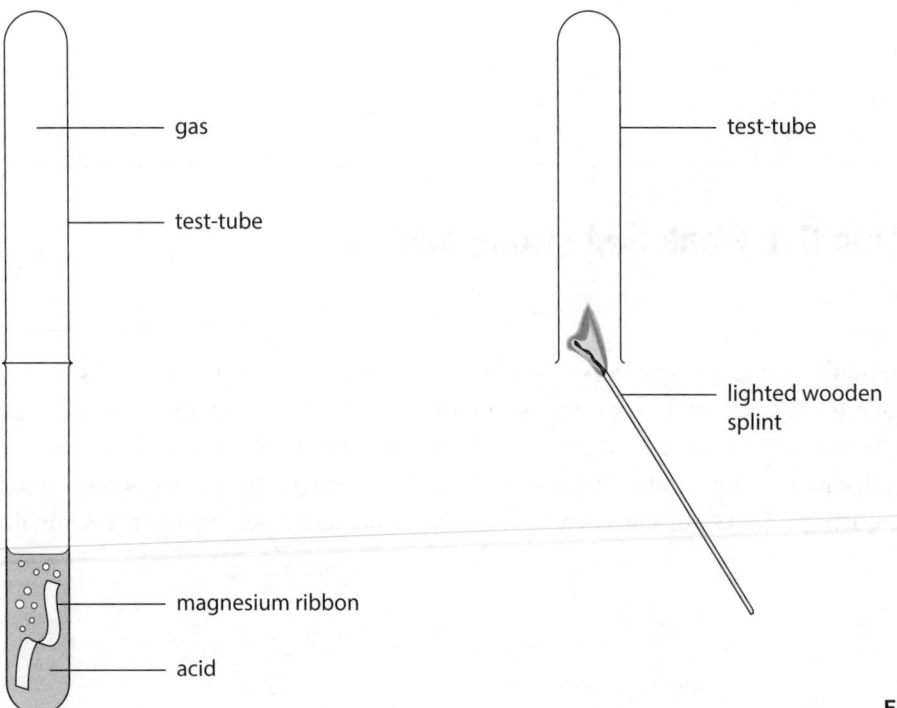

Figure 5.1

7. Add three drops of methyl orange to two dimples on your dropping tile. Rip a small piece of blue litmus paper and add a piece to two dimples.
8. Using a pipette, take a small amount of acid X and add a few drops to the one methyl orange dimple and one litmus paper dimple. Repeat for acid Y. Record your results in the table below.

Safety considerations

Wear eye protection throughout. Report any acid spills to your teacher straight away. The Universal Indicator will stain clothing and skin.

Recording data

Test	Acid X	Acid Y
sodium carbonate – reaction		
sodium carbonate – drops needed		
sodium hydroxide – reaction		
sodium hydroxide – drops needed		
magnesium reaction		
gas given off during magnesium reaction		
blue litmus paper		
methyl orange		
Universal Indicator		

Handling data

On the graph paper provided below, draw a bar graph to show a comparison of the number of drops of sodium carbonate needed to neutralise each acid.

5 Acids, bases and salts

Analysis

1 Complete the sentences below using words from the following list:

> stronger　　neutralise　　metal　　alkalis　　strength　　red
> more　　lighted　　weaker　　vigorous　　hydrogen

In this investigation, various tests were used to compare the .... of the two acids. A strong acid would require .... drops of both sodium hydroxide and sodium carbonate solution to .... them. A .... acid would require fewer drops of the .... to neutralise it. The reaction between strong acids and alkalis is more .... than between weak acids and alkalis. Magnesium is a .... and so will react with an acid to produce .... gas and there will be more effervescence produced in reaction with a .... acid. The test for hydrogen gas is a .... splint, which produces a squeaky pop. Blue litmus paper and methyl orange turn .... in the presence of acid.

2 Use your results to complete these sentences.
Universal Indicator changes .... depending on how acidic or alkaline a solution is. The colour of acid X was .... and therefore it has a pH of .... The colour change observed for acid Y was .... and so it is therefore pH ....

Evaluation

3 Which of the tests you did were useful for determining the strength of each acid?

....

4 Which of the tests were not useful for determining the strength of each acid? Give a reason.

....

....

5 Suggest how many drops of sodium hydroxide would need to be added to an acid with a pH of 4 to neutralise it.

....

Practical investigation 5.2 Reacting acids

Objective

Acids react in similar ways to one another when reacted with other chemicals. In this investigation, you will perform some simple reactions with several different acids. The aim is for you to spot a pattern in the results and then decide on a general rule for the way acids react for each test. By the end of this investigation you should be able to describe the characteristic properties of acids as reactions with metals, bases and carbonates and their effect on litmus and methyl orange.

Equipment

- Five test-tubes
- test-tube rack
- samples of sulfuric
- citric, hydrochloric and ethanoic acid
- methyl orange
- magnesium ribbon
- calcium carbonate
- copper(II) oxide
- spatula
- glass rod
- blue litmus paper
- bung and delivery tube
- limewater

Method

1. Using a pipette, measure 3 cm³ of an acid to each of four of the test-tubes.
2. Add a piece of magnesium ribbon to the first tube. Record your observations.
3. Add half a spatula of copper(II) oxide to the second tube. If nothing happens, you may need to warm the mixture using a hot water bath. Record your observations.
4. Half fill a clean test-tube with limewater. Place the end of the delivery tube in the limewater so that it is submerged (see Figure 5.2). Add half a spatula of calcium carbonate to a test-tube containing acid and then attach the delivery tube and bubble the gas through limewater. Record your observations.

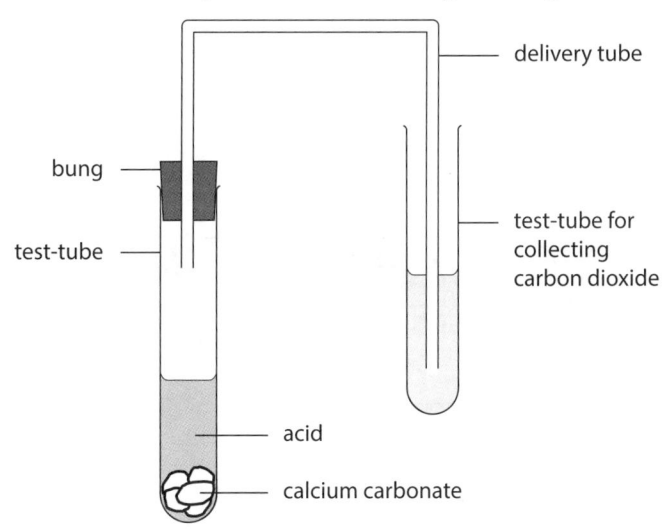

Figure 5.2

5. Dip the glass rod into tube number 4. Carefully touch the glass rod on to a piece of blue litmus paper. Now add a few drops of methyl orange to the test-tube. Observe and record the colour changes.
6. Pour away the samples in all five tubes and rinse them with water. Repeat steps **1–5** with the other three acids.

Safety considerations

Wear eye protection throughout. Report any acid spills to your teacher straight away. Methyl orange is toxic. If you need to use a hot water bath to warm the mixture at step **3** of the method, you will need to stand up. All of the acids and the limewater are irritants. Copper oxide is harmful.

Recording data

In the space below, draw a table to record your results. You will be collecting the results for five different tests for four types of acid.

Analysis

1 Complete the table

Test	General result	General word equation
magnesium		acid + metal →.............................. + ...
copper(II) oxide		acid + base →.............................. + ...
calcium carbonate		acid + carbonate →.................. + ... + ...
blue litmus		
methyl orange		

66 Cambridge IGCSE Chemistry

2 Based on your results, write down the acids in order of strength.

....

Evaluation

3 How do you think your results would be different if you had used copper instead of magnesium as your metal?

....

....

4 Suggest another method you could have used to compare the strength of the acids.

....

Practical investigation 5.3 Reacting alkalis

Objective

Just as with acids, bases react in similar ways to one another when reacted with other chemicals. A dissolved base is called an **alkali**. In this investigation, you will perform some simple reactions with some different alkalis. The aim is for you to spot a pattern in the results and then decide on a general rule for the way alkalis react for each test. By the end of this investigation you should be able to describe the characteristic properties of bases as reactions with acids and with ammonium salts and their effect on litmus and methyl orange.

Equipment

- Sodium hydroxide solution
- calcium hydroxide solution
- six test-tubes
- Universal Indicator
- Universal Indicator chart
- red litmus paper
- ammonium nitrate solution
- methyl orange
- four pipettes
- test-tube rack
- forceps
- glass rod
- beaker (100 cm^3 or 250 cm^3)
- distilled water

Method

1. Using a pipette, add 2 cm^3 of sodium hydroxide solution to three test-tubes and then place these in the test-tube rack.
2. Add a few drops of Universal Indicator to a test-tube containing sodium hydroxide. Record the colour in the results table below. Pour hydrochloric acid into a beaker until you have approximately 0.5 cm depth. Using a clean pipette, add the hydrochloric acid drop by drop into the test-tube. Stop when the solution has been neutralised. Record the number of drops needed in the results table.
3. Using a pipette, add 2 cm^3 of ammonium nitrate to the next test-tube. Make the red litmus paper damp using distilled water and then hold the paper over the mouth of the test-tube using the forceps. Record the colour change.
4. Add three drops of methyl orange to the third tube of sodium hydroxide. Record the colour change in the table.
5. Repeat steps **1–4** using calcium hydroxide.

Safety considerations

Wear eye protection throughout. Report any alkali spills to your teacher straight away. The alkalis are irritants. Methyl orange is toxic.

Recording data

Test	Sodium hydroxide	Calcium hydroxide
Universal Indicator colour		
hydrochloric acid		
ammonium nitrate / red litmus paper		
methyl orange		

Analysis

1 Complete the following paragraph using the words below:

> neutralised   yellow   ammonium   gas   blue   purple   alkalis   green

Both of the hydroxides caused Universal Indicator to change to a .... colour. This showed us that they are .... When hydrochloric acid was added, the colour changed to .... This shows that the alkalis are .... by acids. When the .... nitrate was added to the alkalis, a reaction took place which caused ammonia .... to be released. Ammonia is alkaline and this is why the damp red litmus paper turned .... When methyl orange was added to the alkalis, the solution turned ....

2 Which of the two alkalis was stronger? What evidence do you have to support your answer?

....

....

Evaluation

3 Which of the tests helped you decide which of the two alkalis was stronger?

....

5 Acids, bases and salts

4 Design an experiment using the equipment listed below that you could carry out to determine which of the following two alkalis is stronger. You may include a diagram.

Equipment:
- Bromothymol blue (this is an indicator which is blue in alkalis but green when neutralised)
- burette and clamp stand
- funnel, beakers
- conical flask
- measuring cylinders
- alkalis (ammonia and potassium hydroxide)
- hydrochloric acid, white tile

....
....
....
....
....
....
....
....
....
....

Practical investigation 5.4 The preparation of soluble salts

Objective

Salts are formed when the hydrogen in an acid is replaced by a metal to form an ionic compound. Most people think of table salt (sodium chloride) when they hear the word 'salt' but there are many different types and they are an important part of our day-to-day lives (e.g. fertilisers and food preservatives). Experimentally, we will be examining two types of salt: soluble and insoluble. Soluble salts are formed by neutralising an acid, and insoluble salts are made by precipitation reactions. You will prepare the soluble salts using two different methods:

Method 1: Acid and solid metal, base or carbonate
This method is used when making a salt from a solid metal, a base or a carbonate. The acid has an excess of the solid added to it until no more will dissolve. Any excess material is filtered out and then the solution is crystallised.

Method 2: Acid and alkali
This method is used when preparing a salt from an acid and an alkali. The acid is titrated into the alkali until neutralisation occurs, then the solution is crystallised (see Practical investigations 2.2 and 5.2). As acids and alkalis can be colourless, it is difficult to see when neutralisation has occurred so an indicator needs to be used. By the end of this investigation you should be able to demonstrate knowledge and understanding of the preparation, separation and purification of salts.

Equipment

Soluble salts method 1:
- Beaker (250 cm^3)
- glass rod, funnel
- filter paper
- Bunsen burner
- tripod
- heat-resistant mat
- gauze
- evaporating basin
- measuring cylinder (50 cm^3)
- spatula
- clamp stand and clamp
- tongs
- hydrochloric acid (1.0 mol/dm^3)
- calcium carbonate

Soluble salts method 2:
- Burette (50 cm^3)
- measuring cylinder (25 cm^3)
- clamp stand and clamp
- white tile
- funnel
- conical flask (100 cm^3)
- Bunsen burner
- tripod
- heat-resistant mat
- gauze
- evaporating basin
- methyl orange indicator
- sodium hydroxide solution (1.0 mol/dm^3)
- hydrochloric acid (1.0 mol/dm^3)

Method

Soluble salts method 1

1. Measure 25 cm^3 of hydrochloric acid and pour it into the beaker.
2. Add two spatulas of calcium carbonate to the beaker and stir.
3. Continue adding until no more of the solid dissolves in the acid.
4. Place the filter paper in the funnel, and clamp the funnel above the evaporating basin.
5. Carefully pour the solution into the funnel and collect the filtrate in the evaporating basin.
6. Set up the Bunsen burner and tripod. Place the evaporating basin on the gauze and then gently heat until the liquid begins to boil. Once you see crystals begin to form, stop heating.
7. Using tongs, remove the evaporating basin from the gauze and place it on the heat-resistant mat. Allow to cool.
8. Place the crystals back into a funnel in a filter paper. Wash the crystals carefully with a little distilled water and then place them between two pieces of filter paper so that they can dry.

Soluble salts method 2
1. Measure 25 cm³ of sodium hydroxide using the measuring cylinder and pour it into the conical flask.
2. Add five drops of methyl orange indicator.
3. Rinse the burette with distilled water and then with hydrochloric acid. Make sure that the burette below the tap is also filled. You may need to shake the burette gently while the tap is open to remove any air bubbles. Close the burette tap and fill it with hydrochloric acid to a whole number near zero. (It can be zero but does not need to be; just make sure you record the exact value.) You can refer to the 'Quick skills guide' at the front of this book for guidance on how to read a burette.
4. Secure the burette in the clamp stand and place the conical flask on a white tile.
5. Slowly add the acid to the alkali a few drops at a time until there is a permanent colour change. You will need to swirl the liquid inside the conical flask after you add each portion of acid.
6. Record the final volume in the burette and calculate the volume you added to neutralise the solution. You now have a known volume of acid needed to neutralise the alkali.
7. Pour away the contents of the conical flask and rinse it out with water. Add 25 cm³ of sodium hydroxide to the conical flask. **Do not add indicator.**
8. Refill the burette and then carefully add the known volume of acid from step **6** to the alkalis.
9. Once you have added the know volume of acid, remove the conical flask and then add 10 cm³ of the solution to the evaporating basin.
10. Set up the Bunsen burner, tripod and heat-resistant mat. Place the evaporating basin on the gauze and heat it until the liquid begins to boil. Once you see crystals forming, turn off the Bunsen burner and allow the basin to cool.

Safety considerations

Eye protection must be worn at all times. Sodium hydroxide and hydrochloric acid are both irritants. Take care when moving hot glassware, and use tongs where appropriate. Remember to remove the heat from the evaporating basin and do not allow it to boil dry. As you will be heating liquids, you will need to stand.

Analysis

1. Write a word equation for the reaction taking place during the preparation of the salt by method 1.

   ....

2. Write the balanced symbol equation for this reaction.

   ....

3. Write a word equation for the reaction taking place during the preparation of the salt by method 2.

   ....

4. Write the balanced symbol equation for this reaction.

   ....

5. Describe the colour change you saw in the conical flask during the titration when indicator was present.

   ....

Evaluation

6 Why was the indicator not added to the conical flask when preparing the salts during the second titration?

   ....

   ....

7 Why was the solution formed by reacting calcium carbonate with hydrochloric acid filtered before it was evaporated?

   ....

   ....

8 Describe how you could improve the method for the titration to make sure the volume of acid needed is accurate.

   ....

   ....

Practical investigation 5.5 The preparation of insoluble salts

Objective

An insoluble salt can be prepared using the precipitation method. A precipitate is a solid formed from two solutions during a chemical reaction. In this practical, you will prepare a variety of different salts. The precipitates formed can also be used as a method for identifying unknown solutions. By the end of this investigation you should be able to demonstrate knowledge and understanding of the preparation of insoluble salts by precipitation.

Equipment

- Four test-tubes
- test-tube rack
- glass rod
- distilled water
- five pipettes
- sodium carbonate
- silver nitrate
- copper(II) sulfate
- potassium iodide

Method

1. Rinse each test-tube with distilled water and then place in the test-tube rack.
2. Add about 3 cm³ of sodium carbonate to each of the four tubes.
3. Add about 3 cm³ of silver nitrate to the first tube. Stir using the glass rod. Record your observations in the table.
4. Add about 3 cm³ of each of the other solutions to the remaining tubes of sodium carbonate and record in the results table. Remember to rinse the glass rod each time and use a fresh pipette.
5. Pour away the contents of the test-tubes and rinse them out with distilled water.
6. Repeat steps 2–5 with each of the other solutions so that every combination possible is completed on the results table.

Safety considerations

Wear eye protection at all times. Sodium carbonate is an irritant. Silver nitrate is corrosive. Copper sulfate and potassium iodide are harmful.

Recording data

Fill in the table with your observations (e.g. 'white precipitate formed' or 'no reaction').

	Sodium carbonate	Silver nitrate	Copper sulfate	Potassium iodide
Sodium carbonate				
Silver nitrate				
Copper sulfate				
Potassium iodide				

Analysis

1 For each reaction that produced a precipitate, write the word equation and underline the precipitate that was visible.

....

....

....

....

....

....

....

....

2 List the combinations where no precipitate was formed and suggest a reason why.

....

....

Evaluation

3 Why was distilled water used instead of ordinary tap water?

....

4 Can you suggest a reason for using dilute silver nitrate in this investigation?

....

....

Practical investigation 5.6 Acid titration

Objective

It is sometimes necessary to calculate the concentration of an unknown acid or alkali. In this investigation, you will be given some hydrochloric acid but you will not be told what the concentration is. You will need to use the titration method and some simple calculations to determine the concentration. The titration method is used to find out what volume of an acid is needed to neutralise a known volume and concentration of alkali. It is very important that the readings taken are accurate as they are needed for the calculations and even a small error could have a big impact on the final value. By the end of this investigation you should be able to describe the characteristic properties of acids and their effect on methyl orange.

Equipment

- Burette (50 cm^3)
- pipette or measuring cylinder (25 cm^3)
- clamp stand and burette clamp
- white tile
- funnel
- conical flask (100 cm^3 or 250 cm^3)
- methyl orange indicator
- sodium hydroxide solution (1.0 mol/dm^3)
- hydrochloric acid

Method

1. Measure 25 cm^3 of sodium hydroxide using the measuring cylinder and pour it into the conical flask.
2. Add five drops of methyl orange indicator.
3. Rinse the burette with distilled water and then with hydrochloric acid. Make sure that the burette below the tap is also filled. You may need to shake the burette gently while the tap is open to remove any air bubbles. Close the burette tap and fill it with hydrochloric acid to a whole number near zero. (It can be zero but does not need to be so long as you record the exact value.) You can refer to the 'Quick skills guide' at the front of this book for guidance on how to read a burette.
4. Secure the burette in the clamp stand and place the conical flask on a white tile underneath it (see Figure 5.3).

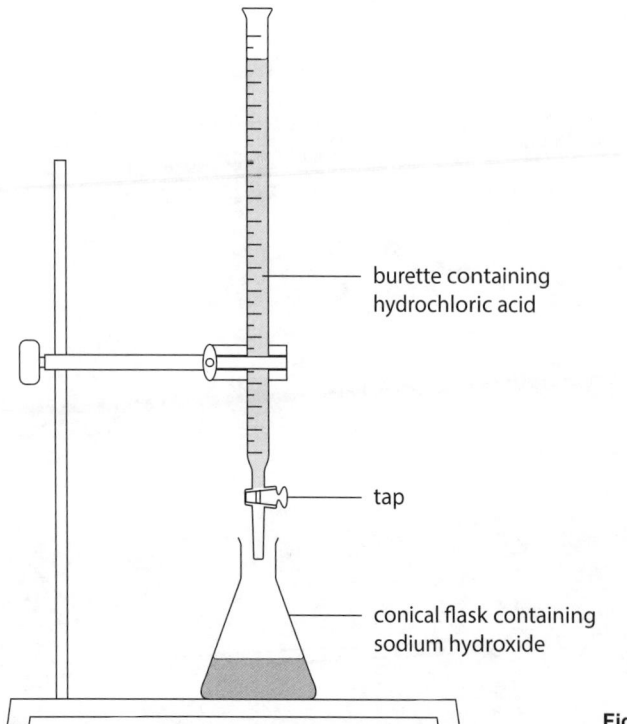

Figure 5.3

76 Cambridge IGCSE Chemistry

5   Slowly add the acid to the alkali a few cm³ at a time until there is a permanent colour change. You will need to swirl the liquid inside the conical flask after you add each portion of acid.
6   Record the final volume in the burette.
7   Calculate the volume that was added to cause the change. This is your rough value.
8   Pour away the contents of the conical flask and rinse it out with water. Repeat steps **1–4**.
9   This time you have a rough value at which the colour change will occur so you can add the acid steadily a few cm³ at a time until you reach a few cm³ less than you did for your rough value and then slowly add the acid to get a more accurate value. Repeat the titration three times and then calculate the mean.

Safety considerations

Wear eye protection throughout. Sodium hydroxide and hydrochloric acid are both irritants. Methyl orange is toxic.

Recording data

Think carefully about how you will record your results for this titration. How many decimal places will you record your data to. The more decimal places the more precise your data will be. Look at your burette carefully. What is the smallest scale division? This will be different for various types of burette but it should be at least to 0.1 cm³.

	Rough value	1	2	3
Final reading / cm³				
Initial reading / cm³				
Volume of acid used / cm³				

Handling data

You can now use the mean volume used to calculate the concentration of the hydrochloric acid.

1   Calculate the number of moles of sodium hydroxide there were in the conical flask.

   ....

   ....

2   How many moles of hydrochloric acid would be needed to neutralise this many moles of sodium hydroxide?

   ....

   ....

3   Calculate the mean volume of hydrochloric acid needed to neutralise the sodium hydroxide. (Remember to ignore your rough results and any anomalous results.)

   ....

   ....

Analysis

4 What was the concentration of hydrochloric acid used?

....

....

Evaluation

5 Why did you ignore the rough results when calculating your mean?

....

....

6 What does 'anomalous' mean?

....

7 Why was it necessary to swirl the conical flask after adding each portion of acid?

....

....

8 Why did you use a white tile?

....

....

9 Were there any problems with your investigation that meant your data might have been unreliable?

....

....

10 Can you suggest a way to improve your investigation so that your results were more reliable?

....

....

Practical investigation 5.7 The pH of oxides

Objective

Oxides are formed when a metal or non-metal is reacted with oxygen. In this investigation, you will attempt to determine whether different types of oxides have different pH values and if there is any pattern to the type of oxides formed by metals and non-metals. By the end of this investigation you should be able to classify oxides as acidic or basic, related to metallic and non-metallic character.

Equipment

Universal Indicator solution, Universal Indicator colour chart, pipettes, six test-tubes, test-tube rack, samples of the following solutions: carbon dioxide, sodium oxide, sulfur oxide, phosphorus oxide, potassium oxide, calcium oxide

Method

You need to find out the pH of six different samples. You have the equipment list above. Plan an investigation in the space below. There is also a diagram for you to label with what you will add to each test-tube (Figure 5.4).

....
....
....
....
....
....
....

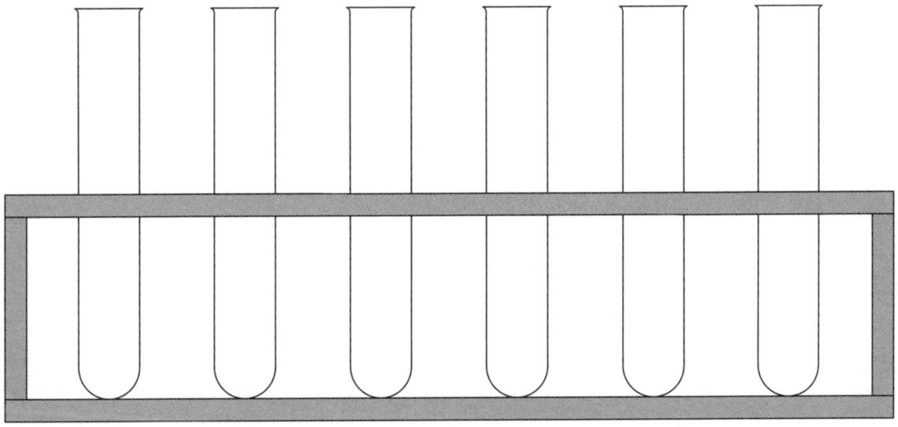

Figure 5.4

Safety considerations

Wear eye protection throughout. Sodium oxide (sodium hydroxide), potassium oxide (potassium hydroxide) and calcium oxide (calcium hydroxide) are all irritants.

Recording data

Name of oxide	Metal/Non-metal	Colour	pH

Analysis

1 Which oxides were acidic?

....

....

2 Which oxides were basic?

....

....

3 Was there a pattern to the type of oxide formed and the pH?

....

....

Evaluation

4 Apart from Universal Indicator, how else could you have measured the pH of each sample?

....

....

Exam-style questions

1   Yousef and Heather are making copper sulfate crystals to be used as a fertiliser for plants. To prepare the salt they must first add excess copper oxide with 25 cm³ of warm sulfuric acid.

   a Suggest suitable apparatus for the following:
      i   To measure the volume of sulfuric acid needed [1]

      ....

      ii To add the copper oxide to the sulfuric acid [1]

      ....

      iii To stir the mixture [1]

      ....

      iv To heat the beaker containing the mixture [1]

      ....

   b What does the word 'excess' mean? [1]

      ....

   c Suggest a method for obtaining the copper sulfate from the solution formed. [5]

      ....

      ....

      ....

      ....

                                                                                                                                Total [10]

2   An investigation into the neutralisation of potassium hydroxide by an unknown acid was conducted by Kwame and Yara. They were trying to find out the temperature change that occurred when the reaction took place. They set up a polystyrene cup and thermometer to measure the temperature change. They measured 25 cm³ of potassium hydroxide and added it to a cup. They added 5 cm³ of the acid at a time, and the temperature of the mixture was recorded.

   a Suggest a suitable piece of apparatus to use to add the acid. [1]

      ....

5 Acids, bases and salts

b Complete the table by reading the temperature from the thermometer diagrams. [3]

Volume of acid added / cm³	Thermometer diagram	Temperature / °C
0	20	
5	23	
10	25	
15	26	
20	27	
25	28	
30	26	
35	25	
40	24	
45	23	
50	22	

c Plot the results of the investigation on the grid below. Add a best-fit line. [4]

d Use your graph to determine the temperature if 13 cm³ of acid were added. [2]

....

e What volume of acid produced the largest increase in temperature? [2]

....

Total [12]

3 Indicators are important because they can be used to determine whether a substance is an acid or a base. Red cabbage can be used to make indicators. Suggest a method for preparing a solution from red cabbage that could be used as an indicator. Include details of how you would check to make sure that the indicator works properly. [5]

....

....

....

....

....

....

Total [5]

5 Acids, bases and salts

6 Quantitative chemistry

In this chapter, you will complete investigations on:

- ◆ 6.1 Determination of the relative atomic mass of magnesium
- ◆ 6.2 Calculating the empirical formula of hydrated salts
- ◆ 6.3 Calculating percentage yield using copper carbonate
- ◆ 6.4 Calculating the concentration of acid using the titration method

Practical investigation 6.1 Determination of the relative atomic mass of magnesium

Objective

The **relative atomic mass** is the average mass of an atom of that element. The aim of this investigation is to calculate the relative atomic mass of magnesium. To do this, the mass of magnesium and volume of hydrogen gas produced must first be obtained when it reacts with an acid. By the end of this investigation you should be able to define relative atomic mass.

Equipment

- Two beakers (100 cm³ and 250 cm³)
- burette
- clamp stand and burette clamp
- small funnel
- emery paper
- magnesium ribbon
- dilute hydrochloric acid
- distilled water

Method

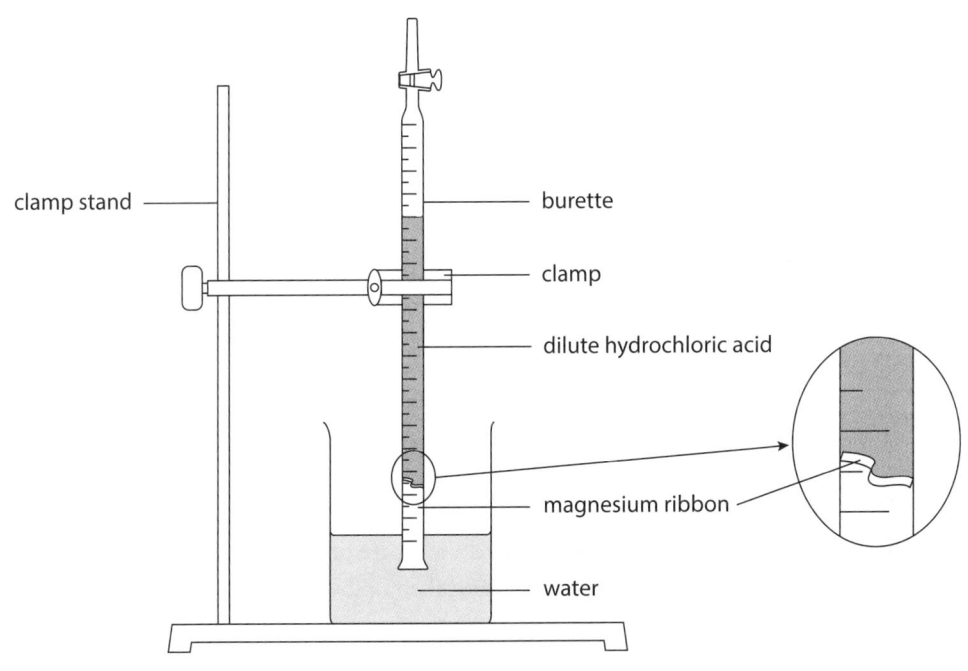

Figure 6.1

1. Using the emery paper, clean the magnesium ribbon so that it is shiny. Weigh the piece of magnesium ribbon and record your result to three decimal places.
2. Using a funnel, carefully pour 25 cm³ of hydrochloric acid into the burette. Now add 25 cm³ of distilled water to the burette. This needs to be done very slowly so that the mixing of liquids is limited.
3. Press the magnesium ribbon into the top of the burette. This should be done widthways so that the magnesium bends and is pressed against the sides of the burette glass. It should not touch the water and should be stuck in place.
4. Half fill the 250 cm³ beaker with distilled water. Quickly invert the burette and place the open end under the surface of the water in the beaker (Figure 6.1). Clamp it in place to the stand and take a reading of the water level.
5. Allow the acid and magnesium to react. Once the liquid level remains the same and no more gas is being produced, the reaction is over. Record the level of liquid in the burette.

Safety considerations

Wear eye protection throughout. Report any spills to your teacher. Hydrochloric acid is an irritant at the concentration you are using.

Recording data

Mass of magnesium .... g

Record your data in the table below.

Burette reading at start / cm³	Burette reading at end / cm³	Volume of gas produced / cm³

Handling data

Using your data, calculate the number of moles of magnesium you had.

1 mole of gas will occupy approximately 24 000 cm³ at 25 °C. To calculate how many moles of hydrogen you have, divide the volume of gas produced by 24 000.

Volume of hydrogen gas produced .... cm³ / 24 000 = ...

Number of moles of hydrogen gas .... mol

The formula for calculating relative atomic mass is given below. In the space provided add your own data to calculate the relative atomic mass of magnesium.

$$\frac{\text{Mass of magnesium}}{\text{number of moles of hydrogen}} = \text{relative atomic mass of magnesium}$$

$$\frac{\text{...............g}}{\text{...............mol}} = \text{...............g/mol}$$

Evaluation

1. To calculate the number of moles of hydrogen gas produced, you used the figure of 24 000 cm³ for the volume of gas occupied by 1 mole of a gas. Which two variables affect the volume taken up by a mole of gas?

   ....
   ....

2. How could you have taken readings to take account of these variables?

   ....
   ....

3. How could you have improved your experiment to make your results more accurate?

   ....
   ....

4. How could you have made your results more reliable?

   ....
   ....

Practical investigation 6.2 Calculating the empirical formula of hydrated salts

Objective

As the mass of water in a hydrated salt is in a fixed proportion to the total mass, by removing water from a hydrated salt it is possible to calculate the proportion of the salt made up by water. By the end of this investigation you should be able to calculate empirical formula.

Equipment

- Two crucibles with crucible lids
- tongs, tripod
- clay triangle
- Bunsen burner
- heat-resistant mat
- copper(II) sulfate
- magnesium sulfate
- two watchglasses
- spatula

Method

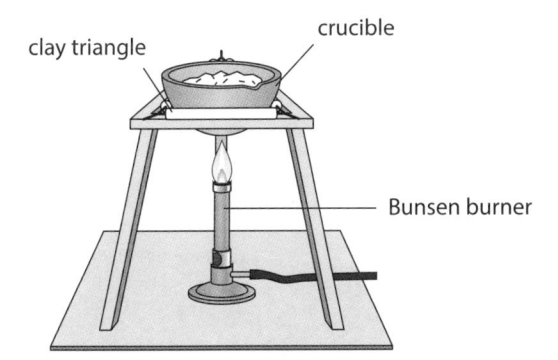

Figure 6.2

1. Take one of the crucibles and weigh it on the balance. Add four spatulas of magnesium sulfate and reweigh. Record the data in the table.
2. Set up the tripod and clay triangle (see Figure 6.2). Place the crucible in the clay triangle and heat the crucible on a blue Bunsen flame for around 5 minutes. The lid can be used to cover the crucible temporarily if the magnesium sulfate starts to spit. The lid should be removed once the spitting stops.
3. Remove the crucible from the clay triangle using the tongs and place it on the heat-resistant mat. Allow it to cool but place a watchglass on top to prevent the reabsorption of water.
4. While waiting for the first crucible to cool, repeat steps **1–3** with the copper(II) sulfate.
5. Return the crucible with magnesium sulfate to the clay triangle and heat a second time but only for 3 minutes. Remove from the heat and allow it to cool. Repeat this step with the copper(II) sulfate.
6. Reweigh both crucibles. Repeat step **5** until a constant mass is achieved – there are extra rows in the table where you can record this data. Your teacher might suggest the use of a desiccator to dry the salts instead of repeated heating.

Safety considerations

Wear eye protection throughout. Do not leave the investigation unattended while the Bunsen burner is on. Remember to use the safety flame when not actively using the Bunsen burner. The crucible will get very hot so use the tongs to move it and allow time for it to cool down before you take measurements.

Recording data

	Magnesium sulfate	Copper(II) sulfate
mass of crucible / g		
mass of crucible and salt / g		
mass of crucible and salt after heating first time / g		
mass of crucible and salt after heating second time / g		
mass of crucible and salt after heating repeat reading / g		
mass of crucible and salt after heating repeat reading / g		
constant mass of crucible and salt / g		

Handling data

Mass of magnesium sulfate at start .... g
Mass of magnesium sulfate at end .... g
Mass of water lost .... g

Mass of copper(II) sulfate at start .... g
Mass of copper(II) sulfate at end .... g
Mass of water lost .... g

Analysis

1. The chemical formula for anhydrous magnesium sulfate is $MgSO_4$. How many moles of magnesium sulfate did you have at the end of the investigation?
   (A_r: O = 16, Mg = 24, S = 32)

   ....

   ....

2. How many moles of water were lost from the magnesium sulfate during the investigation?
   (A_r: H = 1, O = 16)

   ....

   ....

Cambridge IGCSE Chemistry

3 What was the ratio of magnesium sulfate to water?

....

4 The chemical formula for anhydrous copper sulfate is $CuSO_4$. How many moles of copper sulfate did you have at the end of the investigation?
(A_r: O = 16, S = 32, Cu = 64)

....

....

5 How many moles of water were lost from the copper sulfate during the investigation?
(A_r: H = 1, O = 16)

....

6 What was the ratio of copper sulfate to water?

....

Evaluation

7 Can you identify any sources of error in this investigation and suggest improvements?

....

....

....

....

Practical investigation 6.3 Calculating percentage yield using copper carbonate

Objective

The total amount of product of as reaction is not always the same as that predicted by a chemical equation. This is because of a number of factors: the reaction may not be complete, there may be experimental errors (when weighing masses of reactants, for example), and materials may be lost while carrying out the reaction. The comparison between the amount of product that is produced in theory and in reality is called the **percentage yield**. By the end of this investigation you should be able to compare the theoretical and real values for the products produced by the decomposition of copper carbonate and calculate the percentage yield.

Equipment

- Crucible
- tripod
- clay triangle
- Bunsen burner
- heat-resistant mat
- copper carbonate
- tongs, spatula

Method

1. Weigh the crucible and record this value in your table.
2. Add five spatulas of copper carbonate to the crucible and then reweigh.
3. Set up the tripod and clay triangle on the heat-resistant mat.
4. Place the crucible in the clay triangle and then heat using the Bunsen burner on a blue flame.
5. Observe the copper carbonate. When all of it has changed colour from green to black, stop heating.
6. Using the spatula, carefully stir the powder to make sure there is no green colour left. If there is, heat it for a few more minutes.
7. Using the tongs, move the crucible to the heat-resistant mat and allow it to cool.
8. Once the crucible has cooled, reweigh it on the balance.

Safety considerations

Wear eye protection throughout. Copper carbonate is harmful if swallowed. The crucible will get very hot so take care to allow it adequate time to cool before you touch it.

Recording data

Design a table to record the results of your investigation. To get more reliable data, it would be good to include results from at least three other groups.

Handling data

Mass of crucible ....g
Mass of crucible and copper carbonate ....g
Mass of copper carbonate at the start ....g (= mass of crucible and copper carbonate − mass of crucible)
Mass of crucible and copper oxide at the end ....g
Mass of copper oxide at the end ....g (= mass of crucible and copper oxide at the end − mass of crucible)

Analysis

1. What is the molar mass of copper carbonate ($CuCO_3$)?
   (A_r: C = 12, O = 16, Cu = 64)

   ....

2. What is the molar mass of copper oxide (CuO)?
   (A_r: O = 16, Cu = 64)

   ....

3 Calculate how many moles of copper carbonate you had by dividing the mass you measured by the molar mass.
Actual mass / molar mass = number of moles

$$\frac{\text{...................g}}{\text{...................g}} = \text{...................mol}$$

4 The number of moles of copper oxide produced should be the same as the number of moles of copper carbonate reacted. Calculate the expected yield by multiplying the number of moles of copper carbonate you reacted by the molar mass of copper oxide.

Number of moles of copper carbonate × molar mass of copper oxide = expected yield

.....................mol × ....g = ...g

5 You now need to calculate percentage yield. To do this, you must divide your actual yield (the recorded mass of copper oxide) by the expected yield.

(Actual yield / expected yield) × 100 = percentage yield

$$\frac{\text{...................g}}{\text{...................g}} \times 100 = \text{...................\%}$$

Evaluation

6 What was the difference between the expected and actual yield?
.....................%

7 Can you suggest some reasons for this difference?

....

....

....

Practical investigation 6.4 Calculating the concentration of acid using the titration method

Objective

It is important to be able to experimentally calculate concentration of acids and alkalis. This can be done using the **titration method**. As long as the volume and concentration of one of the solutions is known, and the volume needed for neutralisation it is possible to calculate the concentration of the other. In this experiment you will neutralise sulfuric acid using potassium hydroxide. You will have a known volume and concentration of potassium hydroxide. Your aim is to calculate the concentration of the sulfuric acid. By the end of this investigation you should be able to calculate concentrations of solutions.

Equipment

- Clamp stand and burette clamp, burette ($50\,cm^3$)
- conical flask ($250\,cm^3$)
- measuring cylinder ($25\,cm^3$) or pipette ($25\,cm^3$) with pipette filler
- methyl orange
- potassium hydroxide ($0.5\,mol/dm^3$)
- sulfuric acid
- funnel
- beaker ($100–250\,cm^3$)
- methyl orange
- white tile or white paper

Method

1. Using the funnel, add a small amount of the potassium hydroxide to the burette. Place a beaker under the burette. Open the tap and allow the alkali to flow out of the bottom to fill the tip of the burette, then close the tap. If there are any air bubbles, you many need to tap or shake the burette gently to remove them.
2. Pour away the alkali from the beaker and then fill the burette again. The burette needs to be close to the zero mark. Record the reading in your results table. If you need help on how to read the values from a burette, look at the 'Quick skills guide' at the front of this book.
3. Use the measuring cylinder or pipette to measure $20\,cm^3$ of sulfuric acid and then add it to the conical flask.
4. Add three drops of the methyl orange indicator to the conical flask and swirl the contents. Place the conical flask on the white tile or paper underneath the burette.
5. Slowly add a few drops of the potassium hydroxide at a time to the sulfuric acid. After each addition, swirl the conical flask and look for a colour change.
6. When the colour change occurs, stop adding alkali and record the reading on the burette.
7. Pour the contents of the conical flask away and rinse out the conical flask. Repeat steps **1–6** twice more.

Safety considerations

Wear eye protection throughout. Potassium hydroxide is corrosive. Sulfuric acid is an irritant. Methyl orange is toxic.

Recording data

You need to design a results table to write your results into. You will need to think about how many columns and rows you will need.

Handling data

1. Calculate the mean volume of potassium hydroxide needed to neutralise the sulfuric acid (remember to ignore your rough results from the first titration and any anomalous results)

   ....

   ....

2. Write the word equation for this reaction.

   ....

3. Write the balanced symbol equation.

   ....

   ....

Analysis

4 Calculate the number of moles of potassium hydroxide were needed to neutralise the sulfuric acid:

Molarity of potassium hydroxide = ...

Mean volume of potassium hydroxide used = ...

Number of moles = ...

5 How many moles of sulfuric can be neutralised by this many moles of potassium hydroxide?

....

6 What was the concentration of sulfuric acid used?

....

Evaluation

7 Why was it necessary to swirl the conical flask after adding each portion of alkali?

....

....

8 Why did you use a white tile?

....

....

9 Suggest why you were asked to complete a 'rough' titration first.

....

....

10 Can you suggest a way to improve your investigation so that your results were more reliable?

....

....

Exam-style questions

1 A student wanted to find out what products were made when he heated hydrated copper sulfate crystals. He set up the apparatus shown in Figure 6.3.

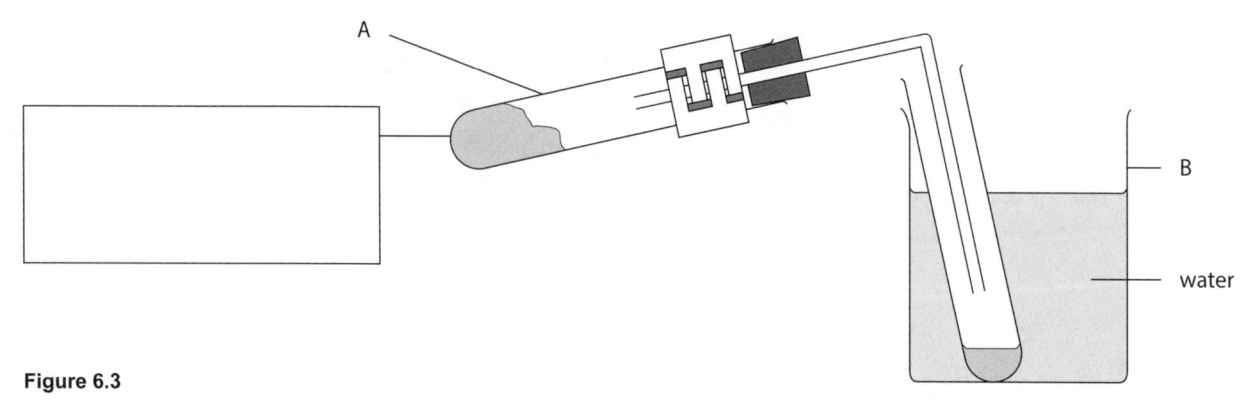

Figure 6.3

    a Look at the diagram and identify the apparatus. [2]
       A ....
       B ....
    b Fill in the box with the name of the material in the boiling tube. [1]
    c Draw an arrow on the diagram to show where heat was applied. [1]
    d The student thought that the liquid that was collected in the test-tube was water. Suggest a way to test if the student was correct. [1]

....

Total [5]

2 Faraaz and Bridgette were investigating a sample of 0.5 mol/dm³ sulfuric acid. They conducted two different titrations with two different concentrations of sodium hydroxide, A and B. The method that they followed is written below.

Using a measuring cylinder, 20 cm³ of sulfuric acid was added to a conical flask. Three drops of methyl orange indicator were added to the acid. A burette was filled to the 0.0 cm³ mark with the sodium hydroxide solution (A or B) and this was added slowly to a conical flask. Once the solution changed colour, the students stopped adding sodium hydroxide. The readings of the burette for each solution are shown in Figure 6.4.

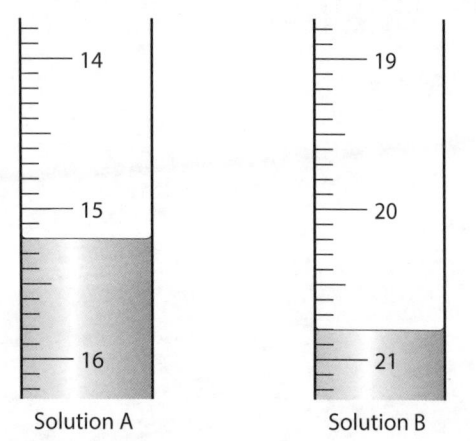

Solution A   Solution B   **Figure 6.4**

a Draw a table for the results. [3]

b Add the readings for solutions A and B to your results table [2]

c Name the type of reaction taking place when sulfuric acid and sodium hydroxide are mixed. [1]

....

d Which solution required more to be added to change the colour of the solution? [1]

....

e What conclusion can you draw from this about the concentrations of solutions A and B? [1]

....

f Using the data in your results table, predict the volume of solution A that would be needed to neutralise 20 cm³ of sulfuric acid at a 1.0 mol/dm³ concentration. [2]

....

g Describe the colour change the students will have seen as the acid is neutralised. [1]

....

Total [12]

3 The shell of a bird's egg is made mostly from calcium carbonate. A student wanted to calculate what percentage of calcium carbonate a bird's egg shell contains. Suggest a method they could use to do this. [9]

....
....
....
....
....
....
....
....
....
....

Total [9]

7 How far? How fast?

In this chapter, you will complete investigations on:

- 7.1 The effect of temperature on reaction rate
- 7.2 The effect of catalysts on reaction rate
- 7.3 Energy changes during displacement reactions
- 7.4 Reversible reactions

Practical investigation 7.1 The effect of temperature on reaction rate

Objective

The rate at which a reaction takes place is affected by a number of variables. If these variables are changed, the reaction can speed up or slow down. There are four main things we can change that can affect the rate of a reaction: temperature, particle size, concentration of reactants, and presence of a catalyst. Temperature affects the rate of reaction because, if particles are at a higher temperature, they have greater kinetic energy and so are more likely to collide successfully. They also collide with greater force and so the activation energy needed for a reaction to take place is more likely to be available. Particle size affects the rate of reaction because the greater the surface available, the more reactions can take place at the same time. Concentration of reactants affects the rate of reaction as there is more likely to be a collision between the reactants if there are a greater number of particles. The presence of a catalyst lowers the activation energy needed for a reaction to take place and therefore increases the rate of reaction.

We can investigate how temperature affects the rate of a reaction by measuring the amount of product produced in a given time. A very simple way to do this is by reacting a solid with a liquid to produce a gas. The gas produced can be measured using a gas syringe. By the end of this investigation you should be able to describe and explain the effect of temperature on the rate of reactions.

Equipment

- Magnesium ribbon
- hydrochloric acid (0.5 mol/dm³)
- conical flask (250 cm³)
- measuring cylinder (100 cm³)
- bung and delivery tube
- gas syringe
- water bath/kettle
- ice
- beaker (250 cm³)
- boiling tube
- thermometer
- clamp stand
- timer

Method

1. Set up the equipment as shown in Figure 7.1.

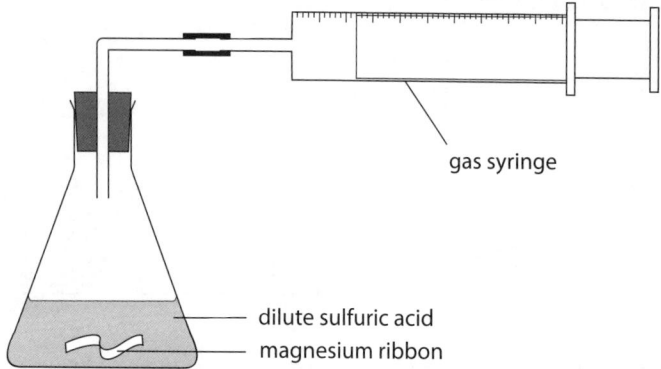

Figure 7.1

2. Measure 25 cm³ of hydrochloric acid using the measuring cylinder and pour it into a boiling tube.
3. Prepare a water bath for the boiling tube. This can be done by adding ice/boiling water to the 250 cm³ beaker. Measure the temperature using the thermometer. You need to use the following approximate temperatures: 0 °C, 10 °C, 15 °C and 30 °C. Remember to add the actual temperature you used to the results table. Place the boiling tube of hydrochloric acid into the beaker until it has reached the required temperature.
4. Pour the hydrochloric acid into the conical flask and quickly add 3 cm of magnesium ribbon. Place the bung back on to the conical flask straight away. Start the timer.
5. In the results table provided, record the volume of gas produced every minute.
6. Once you have filled in all of the data for the first temperature, pour away the acid into the sink and repeat the whole experiment with the next temperature. Do not forget to reset the gas syringe.

Safety considerations

Wear eye protection throughout and report any spills to your teacher straight away. As you will be using hot liquids, you must stand while completing the investigation.

Recording data

Temperature / °C	Volume of gas produced at each time / cm³					
	0 min	1 min	2 min	3 min	4 min	5 min
	0					
	0					
	0					
	0					

Handling data

1   Plot a graph of your results.

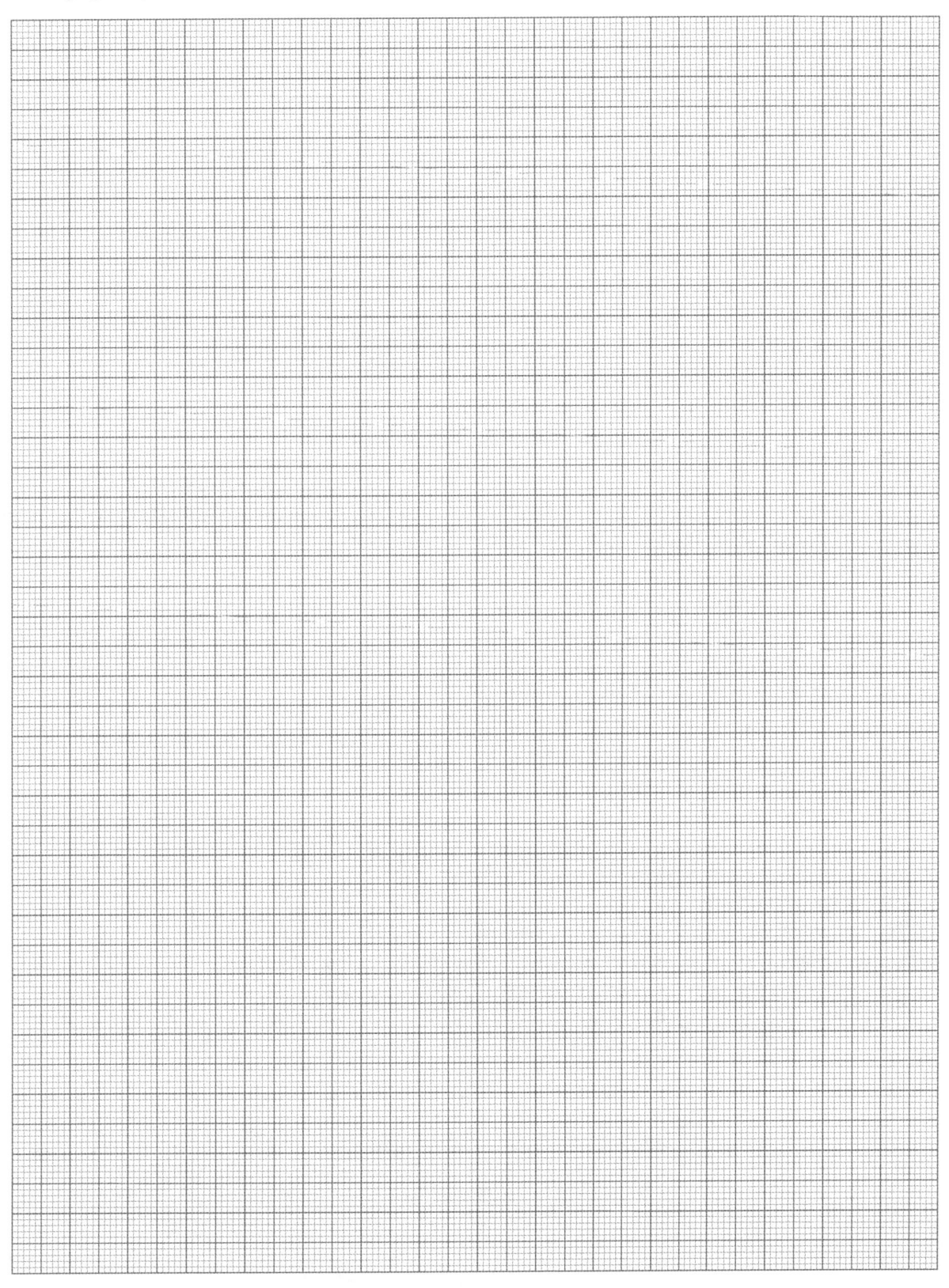

Analysis

2 Look at your graph and use it to find the information to complete the following passage.

As the temperature of the acid is increased, the rate of reaction .... The temperature with the fastest rate of reaction was .... °C. The temperature with the slowest rate of reaction was .... °C. At 30 °C the reaction occurred at its fastest at the start where the line was the .... When the line levelled off, it showed that the reaction was ....

Evaluation

3 What was the independent variable in this experiment – in other words, which variable were you able to change? Circle the correct answer.

Time / Temperature / Concentration of acid

4 What was the dependent variable in this experiment – which variable were you measuring? Circle the correct answer.

Time / Temperature / Volume of gas produced

5 List three variables that were controlled in this experiment.

....

6 If the experiment was repeated with acid heated to 60 °C, how long do you think it would take for the reaction to finish? Give a reason for your answer.

....

....

7 Name the gas that was produced by this reaction.

....

8 How could you test the gas to see if you are correct?

....

9 Read the statement below.

'This experiment had some sources of error that may have affected the results.'

Can you identify two sources of error and suggest what effect they might have had on the results obtained?

Source of error	Effect on results

Cambridge IGCSE Chemistry

Practical investigation 7.2 The effect of catalysts on reaction rate

Objective

One of the variables that affects the rate of a reaction is the presence of a catalyst. In this investigation, you will use a variety of catalysts with the aim of finding which one is best. By the end of this investigation you should be able to describe and explain the effect of catalysts on the rate of reactions.

Equipment

- Timer
- two glass measuring cylinders (25 cm^3)
- dropping pipette
- scrap paper
- black marker
- beaker (250 cm^3)
- samples of catalysts (copper(II) sulfate solution, iron(II) sulfate solution)
- 100 cm^3 sodium thiosulfate solution
- 100 cm^3 iron(III) nitrate solution

Method

1. Using the black marker, draw a thick black cross on the scrap paper. Place this under the beaker. Add 25 cm^3 of sodium thiosulfate solution to the beaker.
2. Look down on the beaker from above. The black cross should be clearly visible (see Figure 7.2).

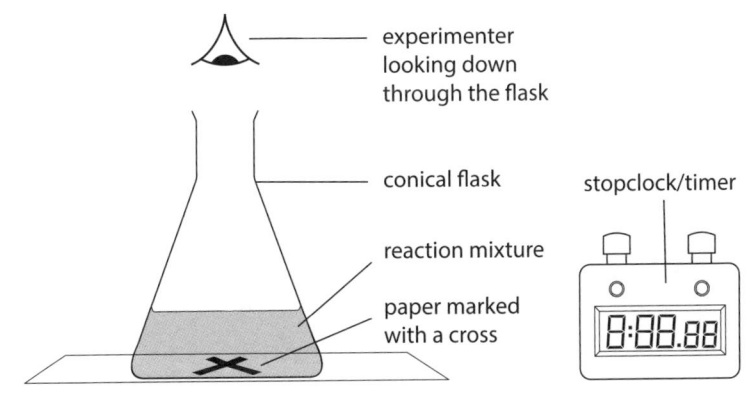

Figure 7.2

3. Measure 25 cm^3 of iron(III) nitrate solution using the other measuring cylinder.
4. Prepare the timer then pour the iron(III) nitrate solution into the beaker. Start the timer.
5. The solution should initially become opaque and the cross should disappear. Stop the timer when the cross can be seen again.
6. Pour away the sodium thiosulfate and iron(III) nitrate mixture and rinse out the beaker.
7. Repeat steps **2–6** but this time add a single drop of one of the catalysts to the iron(III) nitrate while it is still in the measuring cylinder.
8. Repeat for each catalyst and record the data obtained in your results table.

Safety considerations

Wear eye protection throughout.

Recording data

Design a table to show the reaction time for each of the catalysts and for when no catalyst was used.

Handling data

Plot a bar graph to show your results.

Analysis

1 List the catalysts from most effective to least effective.

....

Evaluation

2 In this investigation, only one drop of each catalyst was used. Why?

....

3 Some of the catalysts used in this reaction caused decolourisation in very little time. How could the reaction have been slowed down? (Think about which other variables affect the rate of a reaction.)

....

....

4 Can you suggest a more accurate method for determining when the reaction was over?

....

....

Practical investigation 7.3 Energy changes during displacement reactions

Objective

A more reactive metal will displace a less reactive metal from a solution of its salts. In this investigation, you will measure the energy released when metals of different reactivity are added to a solution. By the end of this investigation you should be able to deduce an order of reactivity of metals from a given set of experimental results.

Equipment

- Polystyrene cup and lid with a hole
- beaker (250 cm^3)
- thermometer, measuring cylinder (25 cm^3 or 50 cm^3)
- copper(II) sulfate solution
- spatula, clamp stand and clamp
- glass rod
- zinc powder
- iron filings
- magnesium powder
- copper powder

Method

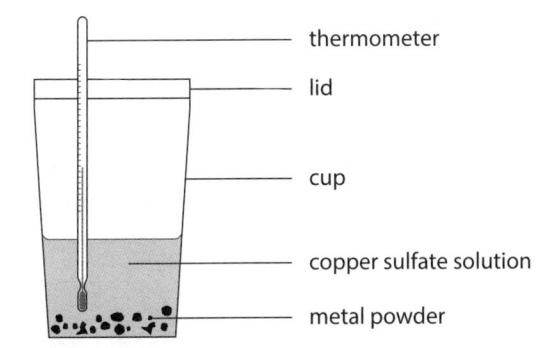

Figure 7.3

1. Design your results table to record your data.
2. Place the polystyrene cup inside the beaker to prevent it from falling over. Attach the thermometer to the clamp on the stand so that you are able to lower it into the polystyrene cup through the lid (see Figure 7.3).
3. Measure 25 cm^3 of copper(II) sulfate and pour it into the cup. Lower the thermometer into the cup and record the temperature.
4. Add one spatula of iron filings to the cup and stir with the glass rod. Record the highest temperature reached. (This may take a few minutes.)
5. Remove the thermometer and pour away the mixture following your teacher's instructions. Rinse the cup out with water. Repeat steps **2–3** with the magnesium, copper and zinc powder.

Safety considerations

Wear eye protection throughout. Copper(II) sulfate is harmful. All of the powdered metals except for copper are flammable.

Recording data

Design a table to show your results. Remember you need enough rows for each type of metal. Add enough columns to record the starting temperature, highest temperature and the temperature change.

Handling data

Plot a graph to show the temperature change for each type of metal.

Analysis

1 Which of the metals produced the biggest temperature change?

....

2 Why do you think this was?

....

3 Which metal produced the smallest temperature change?

....

4 Why do you think this was?

....

5 Can you put the metals into a reactivity series based on the size of the temperature changes observed?

....

....

Evaluation

6 Why did you have to measure the starting temperature of the copper(II) sulfate?

....

....

7 List the variables that you controlled in this investigation.

....

....

8 How could you have obtained more reliable data?

....

....

9 Name a variable that you did not control in this investigation.

....

10 What effect could this have had on your results?

....

....

11 Did your results match the order of the reactivity series?

....

....

Practical investigation 7.4 Reversible reactions

Objective

Some reactions are reversible and will move in different directions (forwards favouring the formation of the products or backwards favouring the formation of the reactants) depending on the conditions present. In this investigation, you will examine three different reversible reactions and how they can be made to move in either direction. By the end of this investigation you should be able to understand that some chemical reactions can be reversed by changing the reaction conditions.

Equipment

Copper ions equilibrium:
- Test-tube rack
- three test-tubes
- two pipettes
- copper(II) sulfate solution
- ammonia solution
- dilute sulfuric acid solution
- measuring cylinder (10 cm^3)

Carbon dioxide and water:
- Sodium hydroxide solution
- conical flask (250 cm^3)
- phenol red indicator solution
- distilled water
- measuring cylinder (100 cm^3)
- tripod
- gauze
- spatula
- Bunsen burner
- heat-resistant mat

Hydrated copper(II) sulfate:
- Copper(II) sulfate
- evaporating basin
- tripod
- gauze
- spatula
- Bunsen burner
- heat-resistant mat
- distilled water

Method

The investigations can be performed in any order.

Copper ions equilibrium
1. Place the test-tubes in the rack and add 1 cm^3 of copper(II) sulfate to the first two tubes.
2. Using the pipette, add ammonia drop-wise to the first tube. Be sure to shake the tube gently after each drop is added. Record the colour change in the results table. Pour half of this mixture into tube 2 and half into tube 3.
3. Continue to add ammonia one drop at a time to tube 2 until you see another colour change.
4. Add dilute sulfuric acid one drop at a time to the tube 3. Remember to shake the tube after adding each drop. When you observe a colour change, stop adding the acid. Record your observations.

Carbon dioxide and water
1. Pour 100 cm^3 of distilled water into the conical flask.
2. Add two drops of phenol red indicator to the water.
3. Add drops of sodium hydroxide to the solution until it turns red.
4. Breathe into the flask until the solution turns yellow.
5. Set up the tripod, gauze, heat-resistant mat and Bunsen burner.
6. Gently heat the conical flask until the colour changes back to red. (Do not boil the solution.)

Hydrated copper(II) sulfate
1. Add four spatulas of copper(II) sulfate to the evaporating dish.
2. Set up the tripod, gauze, heat-resistant mat and Bunsen burner. Place the evaporating basin on the gauze and heat it gently on a blue flame.

3   Stop heating when the copper(II) sulfate changes from blue to white.
4   Allow the evaporating basin to cool back to room temperature for about 5 minutes. You could use this time to pack away the rest of the apparatus or begin one of the other experiments.
5   Add a few drops of distilled water. Record the colour change.

Safety considerations

Wear eye protection throughout. Copper(II) sulfate is harmful. Dilute sulfuric acid and sodium hydroxide are irritants. Phenol red indicator is highly flammable so keep it away from any naked flames. Take care when adding water to the evaporating basin; make sure that you have allowed it to cool first.

Recording data

Copper ions equilibrium

Tube number	Observations
1	
2	
3	

Carbon dioxide and water

	Colour observed
Sodium hydroxide added	
Air exhaled into the flask	
Flask heated	

Hydrated copper(II) sulfate

	Colour observed
After heating	
After adding water	

Analysis

1   What type of reaction is occurring between the ammonia and the sulfuric acid in the copper ions equilibrium reaction?

   ....

2   Why was sodium hydroxide added to the conical flask in the carbon dioxide and water investigations?

   ....

3   Why did heating the conical flask return the solution to an alkaline pH?

   ....

Evaluation

4 Design an experiment to collect the water that was given off by the hydrated copper(II) sulfate when it was heated.

....

....

....

....

Exam-style questions

1 Sultan was investigating the effect of surface area on the rate of reaction. He had three samples of calcium carbonate: powder, small lumps and one large lump. Calcium carbonate reacts with sulfuric acid to produce carbon dioxide gas. The student measured the volume of gas produced each minute for 5 minutes using a gas syringe for each of the three samples.

a How could the student test that the gas produced was carbon dioxide? [1]

....

....

b Suggest two control variables for this experiment. [2]

....

....

c Complete the table below for sulfuric acid. [3]

Test	Methyl orange	Blue litmus paper	Universal Indicator
Colour			

The results of the investigation are given in the table.

Calcium carbonate	Powder	Small lumps	Large lump
Time /	Volume of gas produced /	Volume of gas produced /	Volume of gas produced / ....
0	0	0	0
1	18	9	3
2	25	15	5
3	25	19	6
4	25	23	7
5	25	25	8

7 How far? How fast?

d Add in the missing units to the column headings. [2]

e Plot a graph on the grid below to show the results. [5]

f Suggest how many minutes it took for the all of the powdered calcium carbonate to react, and give a reason for your answer. [2]

....

....

g A second student had a sample of more finely ground calcium carbonate powder. Use the information from the table and your graph to predict how long it would take for this powder to completely react with the acid. [1]

....

Total [16]

2 The reactivity series can be used to predict if one metal will replace another from a solution of one of its salts. This reaction will release energy in the form of heat.

Rabab wanted to investigate the temperature changes that occurred when they added different metals to lead sulfate. They followed the method described below.

Using a pipette, 25 cm³ of lead sulfate were added to a glass beaker. The temperature of the lead sulfate was measured using a glass thermometer. A spatula of magnesium powder was added to the beaker and the mixture was stirred. The highest temperature reached was recorded. This method was repeated with zinc, iron and copper powder.

a Suggest three improvements to the method described above. [3]

....

....

....

....

b Look at the results table below. Add the metal powder that you think caused each temperature change. The starting temperature of the lead sulfate was 24 °C. [4]

Metal powder	Highest temperature reached / °C
	38
	24
	28
	41

Total [7]

3 Hydrogen peroxide decomposes to form oxygen and water. The reaction can be summarised using the following symbol equation:

H_2O_2 .... → H_2O .... + O_2 ....

a Add the state symbols to the symbol equation above. [3]

This reaction happens very slowly. It is possible to increase the speed of this reaction by adding a catalyst.

b Design an experiment to compare the three catalysts iron(III) oxide, manganese(IV) oxide and lead(IV) oxide to see which one has the greatest effect on the speed of reaction. [6]

....

....

....

....

....

....

Total [9]

7 How far? How fast?

8 Patterns and properties of metals

In this chapter, you will complete investigations on:

- 8.1 The extraction of iron
- 8.2 The extraction of copper from malachite
- 8.3 The reactions of metals and acids
- 8.4 Investigating the reactivity series using an electrochemical cell
- 8.5 Using flame tests to identify metals

Practical investigation 8.1 The extraction of iron

Objective

Metals are essential materials for modern life. They are used in everything from electronics to building materials. However, very few metals are found in a pure state; most are in compounds with other elements. Minerals that metals are extracted from are called **ores**. The main ore of iron is called haematite, which contains large amounts of iron oxide. To obtain pure iron, the oxygen must be removed by a more reactive element. In this investigation, you will use carbon to remove the oxygen from the iron oxide. By the end of this investigation you should be able to describe and state the essential reactions in the extraction of iron from hematite.

Equipment

- Sodium carbonate powder
- iron(III) oxide powder
- match
- tongs
- Bunsen burner
- heat-resistant mat
- magnet
- resealable plastic bag
- pestle and mortar
- white paper

Method

1. Moisten the match-head with water.
2. Roll the head of the match in the sodium carbonate powder.
3. Roll the match in the iron(III) oxide powder.
4. Set up the Bunsen burner on the heat-resistant mat.
5. Light the Bunsen burner and set it to a blue flame.
6. Pick up the match with the tongs and hold it in the flame. Allow the match to burn half way along the length before blowing it out.
7. Turn off the Bunsen burner and allow the match to cool on the heat-resistant mat.
8. Place the charred part of the match in the pestle and mortar and grind it for a few seconds so that you have a fine powder.
9. Seal the magnet in the plastic bag and then dip it into the powder. Remove it from the powder and then blow gently to remove any loose material.
10. Hold the bag over white paper and remove the magnet. Look for any pieces that have fallen from the bag.

Safety considerations

Wear eye protection throughout. Sodium carbonate powder is an irritant.

Analysis

1. What were the small black specks that you saw on the white paper when you removed the magnet?

   ....

   ....

2. What was the source of carbon in the reaction?

   ....

   ....

3. Write the word equation for the reaction taking place.

   ....

Evaluation

4. Why was a match used for this investigation?

   ....

   ....

5. What is the name given to this type of reaction?

   ....

Practical investigation 8.2 The extraction of copper from malachite

Objective

Copper is a very valuable metal and is used extensively in electronics. The main ore of copper is called malachite, which is formed of copper carbonate. As copper is already combined with carbon, it first needs to be turned into copper oxide. Once this reaction is complete, carbon can be used to reduce the copper oxide to pure copper. By the end of this investigation you should be able to describe the ease in obtaining metals from their ores by relating the elements to the reactivity series.

Equipment

- Bunsen burner
- heat-resistant mat
- tripod
- clay triangle
- crucible
- spatula
- copper(II) carbonate
- powdered carbon
- beaker (250 cm^3)

Method

1. Set up the tripod with the clay triangle on the heat-resistant mat.
2. Add two spatulas of the copper(II) carbonate powder to the crucible. Place the crucible into the clay triangle.
3. Light the Bunsen burner and carefully heat the crucible on a blue flame. Look for a colour change from green to black.
4. Allow the crucible to cool for 5 minutes.
5. Add three spatulas of carbon powder to the crucible. Use the tongs to hold the edge of the crucible to stop it from moving, and then stir the contents using the spatula. Make sure that the two powders are well mixed.
6. Add one more spatula of carbon powder to the crucible. Carefully sprinkle this over the surface to form a layer on the top.
7. Heat the crucible using a blue flame at first and then on a roaring flame until the reaction mixture glows orange.
8. Turn the Bunsen burner off and allow the crucible to cool.
9. Half fill the beaker with water.
10. Holding the crucible with the tongs, pour the powder from the crucible into the beaker.
11. Stir the contents of the beaker with the spatula and allow the sediment to settle. Carefully pour off the water, which should contain a black suspension of carbon powder. Add more water and repeat this step until only the heavy material is left at the bottom of the beaker.

Safety considerations

Wear eye protection throughout. Copper(II) carbonate is harmful. The crucible will be very hot so grip it with tongs carefully and make sure you do not touch it with your hands.

Analysis

1. Write the word equation for the reaction that took place when you heated the copper carbonate.

   ....

2. What is the formula equation for this reaction?

   ....

3 Write the word equation for the reaction that took place when you heated the copper oxide with the carbon.

....

4 What is the formula equation for this reaction?

....

Evaluation

5 Why did the copper sink to the bottom of the water?

....

....

6 What was the purpose of the carbon powder?

....

....

Practical investigation 8.3 The reactions of metals and acids

Objective

Most metals will react with acids to form a salt. A by-product of this reaction is that a gas is produced. The more reactive a metal is, the more vigorous the reaction will be. If we observe the reaction of a metal and an acid, we can order metals into a reactivity series. By the end of this investigation you should be able to place magnesium, zinc, iron, tin and copper in order of reactivity by reference to the reactions of the metals with dilute hydrochloric acid.

Equipment

- Test-tubes
- test-tube rack
- dilute hydrochloric acid
- dilute sulfuric acid
- spatula
- pipette
- samples of magnesium
- zinc
- iron
- tin and copper

Method

1. For this investigation, you must plan the method and have it checked by the teacher before beginning experimental work. Your aim is to observe the different metals when they react with the acids. Remember to number your steps.

   ....
   ....
   ....
   ....
   ....
   ....
   ....
   ....
   ....
   ....

Safety considerations

2. What safety precautions will you need to take?

   ....
   ....

Recording data

Metal	Reaction with dilute sulfuric acid	Reaction with dilute hydrochloric acid

Analysis

3   Put the metals in order of vigorousness of reaction with sulfuric acid.

   ....

   ....

4   Put the metals in order of vigorousness of reaction with hydrochloric acid.

   ....

   ....

5   Name the gas produced in these reactions.

   ....

6   Suggest a test for the gas.

   ....

   ....

7   Suggest which of these two acids is stronger. What evidence do you have to support your answer?

   ....

   ....

Evaluation

8 The results you have obtained are descriptive. How could you improve the investigation to obtain numerical data?

....
....
....
....

Practical investigation 8.4 Investigating the reactivity series using an electrochemical cell

Objective

When two metals of different reactivity are placed into a circuit with an electrolyte, a small current is generated. Once there is current generated in the cell, it is possible for us to measure the voltage. In this investigation, you are going to use the information about the size of the voltage produced in the electrochemical cell to put the metals used into an order of reactivity. By the end of this investigation you should be able to deduce an order of reactivity from a given set of experimental results.

Equipment

- Sodium chloride solution
- strips of unknown metals
- two wires with crocodile clips
- copper strip
- voltmeter
- two beakers (50 cm^3)
- measuring cylinder (50 cm^3 or 100 cm^3)

Method

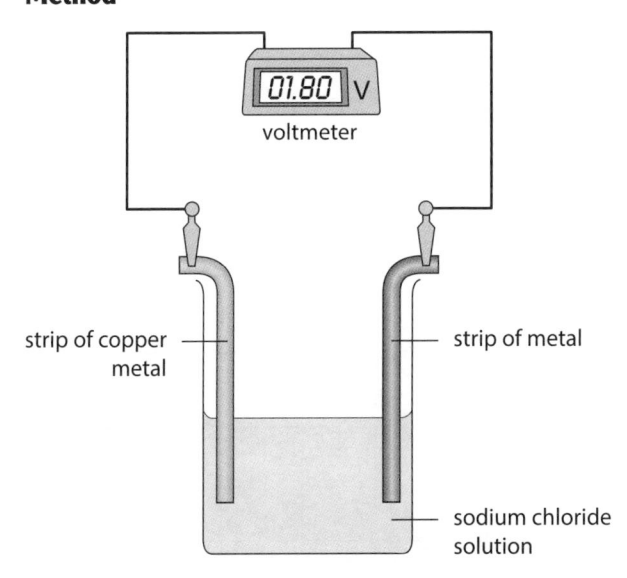

Figure 8.1

1. Connect the voltmeter to the two wires as shown in Figure 8.1.
2. Half fill each beaker with sodium chloride solution.
3. Attach the copper strip to one of the crocodile clips and fold it over the edge of the beaker so that it does not move and so that part of it is submerged in the sodium chloride solution.
4. Take unknown metal A and attach it to the other crocodile clip. Fold it over the edge of the beaker so that it does not move and so that part of it is submerged in the sodium chloride solution.
5. Record the voltage shown on the voltmeter in your results table.
6. Repeat step **4** with each of the unknown metals.

Safety considerations

Wear eye protection throughout.

Recording data

Draw a table to record your results in the space below.

Handling data

On the graph paper provided, draw a bar graph to show your results.

Analysis

1 Based on your results, put the unknown metals in order of reactivity.

....

....

2 Suggest what each of the unknown metals were.
Metal A ....
Metal B ....
Metal C ....
Metal D ....

Evaluation

3 How could you check that the results you obtained for each metal are accurate?

....

....

4 From your results, can you predict the voltage that you would get if you used a strip of magnesium?

....

Practical investigation 8.5 Using flame tests to identify metals

Objective

One way to indentify the metals present in a compound is by observing the colour produced during a flame test. In this investigation, you will be given four different metal compounds. You will need to observe the colour produced by these samples. You will then be given three unknown samples and you will be required to identify them. By the end of this investigation you should be able to use the flame tests to identify lithium, sodium, potassium and copper(II).

Equipment

- Bunsen burner
- distilled water
- seven wooden splints (soaked)
- heat-resistant mat
- test-tube rack
- beaker (250 cm³)
- solutions of the following:
    lithium chloride, sodium chloride, copper chloride, potassium chloride, sample X, sample Y and sample Z

Method

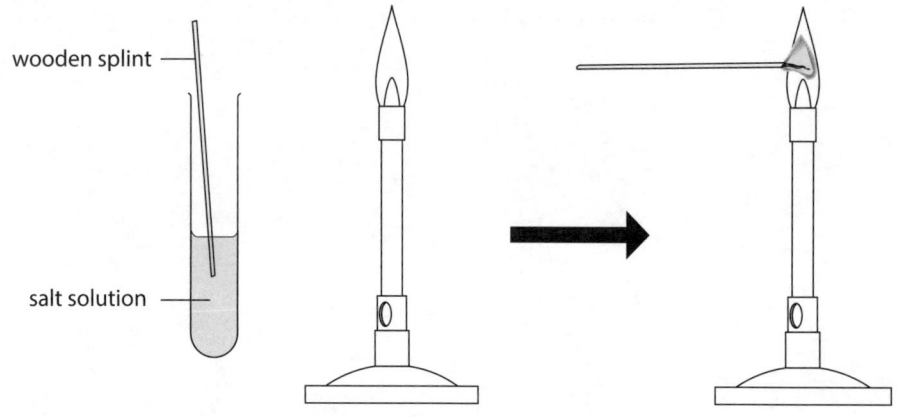

Figure 8.2

1. Place your splint into the solution of lithium chloride and allow it to soak for a few minutes. Half fill the beaker with tap water.
2. Set up the Bunsen burner and heat-resistant mat.
3. Remove the splint from the solution of lithium chloride and heat it in the Bunsen using a gentle blue flame (see Figure 8.2). Record the colour observed in the table.
4. Dispose of your splint in the beaker of water.
5. Repeat steps **1–3** with each of the other samples.

Safety considerations

Wear eye protection at all times. Lithium chloride and copper chloride are harmful. Z is an irritant. Stand up when using the Bunsen burner to heat the splints.

Recording data

Sample	Colour observed
lithium chloride	
sodium chloride	
copper chloride	
potassium chloride	
sample X	
sample Y	
sample Z	

Analysis

1. From your results, suggest which metal was contained in samples X, Y and Z.
   Sample X ....
   Sample Y ....
   Sample Z ....

Evaluation

2. Which unknown sample could you not identify?

   ....

3. Where could you find information to help you identify this sample?

   ....

4. What would a control be for this investigation?

   ....

Exam-style questions

1 Angus and Danni have a sample of copper carbonate. Their teacher has asked them to plan an experiment to get some pure copper from the copper carbonate. They have the following equipment:

> Bunsen burner, boiling tube, carbon powder, spatula, test-tube holder, heat-resistant mat

    a Write a short plan for how they could get pure copper using the apparatus list above. [4]

    b Write word equations for the reactions taking place in your method. [6]

    c The copper obtained from using the reduction method is not pure. Suggest what further steps the students would have to take to obtain pure copper. [3]

    d Suggest two uses of copper. [2]

Total [15]

2 Kendra is investigating the colours produced by different solutions using a flame test.

    a Suggest a safety precaution the student should take while performing the flame tests. [1]

Kendra had three unknown compounds, A, B and C.

Sample	Flame colour	Metal
A	lilac	
B	green	
C	red	

b Add the names of the metals present in each sample to the table. [3]

Total [4]

3 Chris is investigating the reactivity of different metals (see Figure 8.3). He wants to measure how much gas is given off in 1 minute when he adds each metal to 20 cm³ of sulfuric acid. The gas produced during the reaction will be collected in an inverted measuring cylinder. Each metal is carefully cleaned before the experiment.

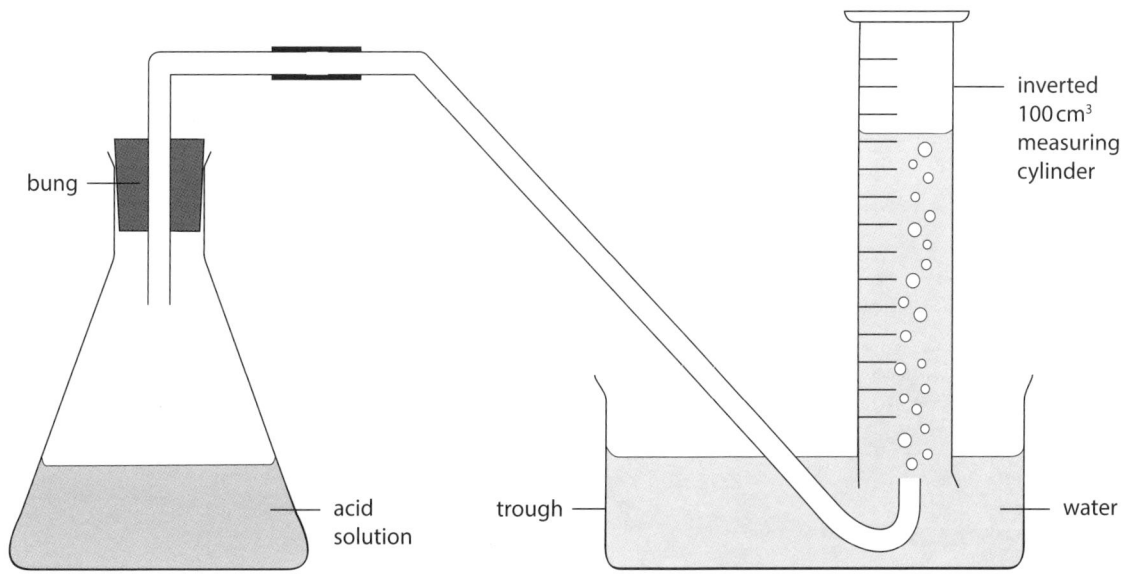

Figure 8.3

a Suggest a piece of apparatus Chris could use to time 1 minute. [1]

....

b Look at the measuring cylinder diagrams shown in Figure 8.4. Draw a table and record the volume of gas produced by each metal in 1 minute. [6]

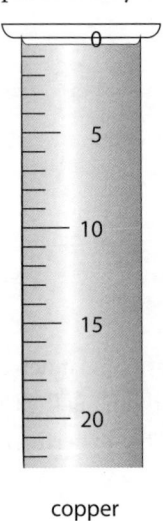

copper

magnesium

aluminium

iron

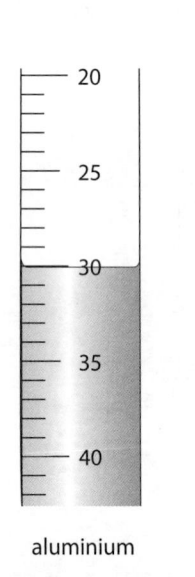

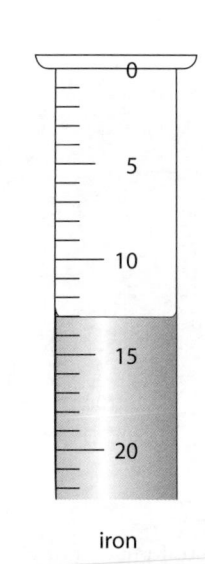

Figure 8.4

c How could Chris have obtained more accurate data for the volume of gas produced? [1]

....

d Why were the metals cleaned before the experiment? [1]

....

130 Cambridge IGCSE Chemistry

e Using the data, plot a graph to show the results on the grid below. [4]

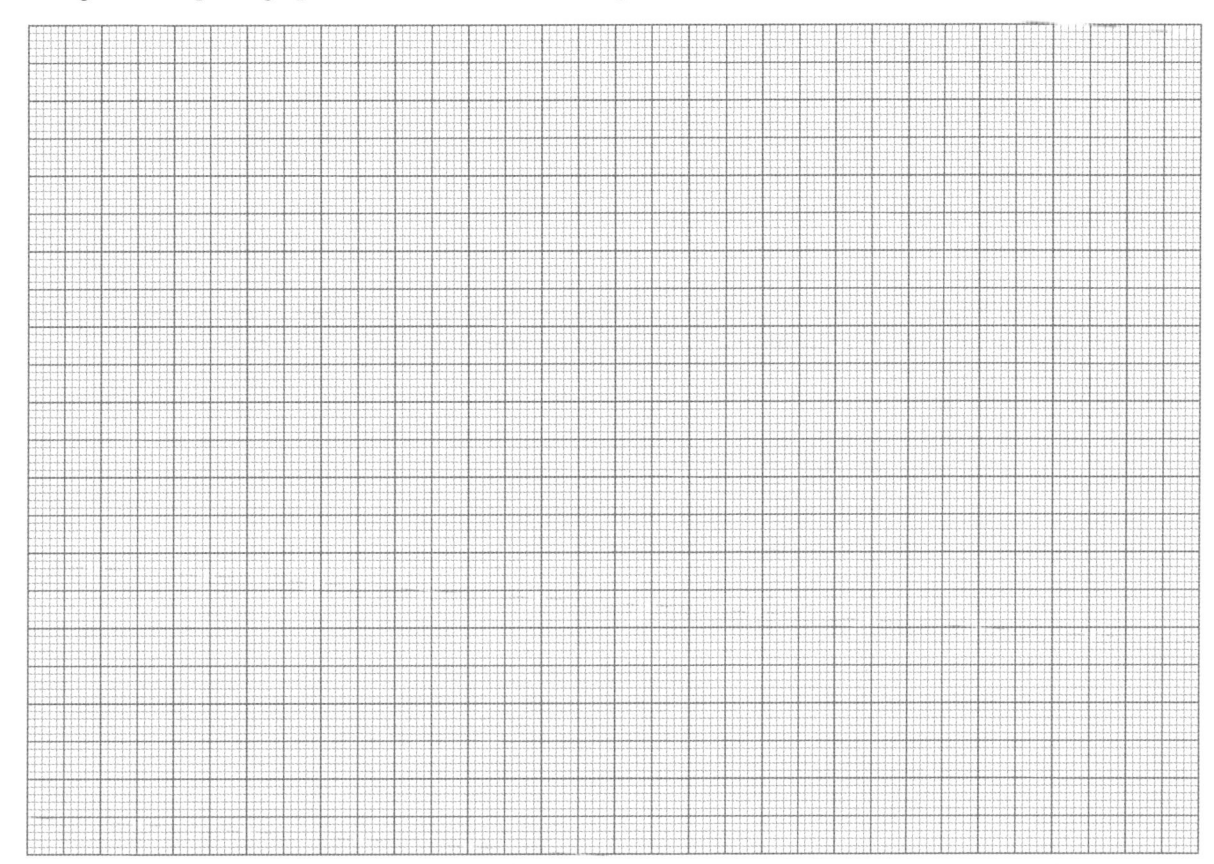

f Using the results table and your graph, put the metals in order of reactivity starting with the most reactive. [3]

....

....

Chris repeated the experiment with an unidentified metal. It produced the result shown in Figure 8.5.

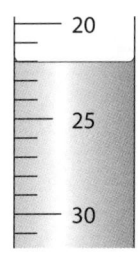

Figure 8.5

g Using the data and your answer to part **f**, suggest which metal this might be with a reason. [2]

....

....

8 Patterns and properties of metals   131

h What is the name of the gas being produced? [1]

....

i How could you test the gas, and what result would you expect to see if your suggestion in **h** is correct? [2]

....
....

j Chris did not control all of the variables in his experiment. Name two variables that he did not control and explain how he could improve his method to make sure that these variables were controlled. [4]

....
....
....
....
....

Total [25]

9 Industrial inorganic chemistry

In this chapter, you will complete investigations on:

◆ 9.1 What causes rusting?

◆ 9.2 Preventing rusting

◆ 9.3 Using limestone (calcium carbonate) to neutralise acidic water

Practical investigation 9.1 What causes rusting?

Objective

In this investigation, you will examine the different factors that affect the chemical reaction commonly called **rusting**. Rusting is a reaction in which iron reacts with oxygen and water to produce hydrated iron oxide, a red-brown powder. Iron and steel are important metals with many uses in our everyday lives, from cars to building materials. Rust causes the breakdown of any object made of iron or steel, and this has serious economic implications. In this investigation, you will need to pay particular attention to controlling variables and also making careful observations. By the end of this investigation you should be able to state the conditions required for the rusting of iron.

Equipment

- Four test-tubes
- four iron nails
- distilled water
- test-tube rack
- boiled water
- oil
- anhydrous calcium chloride
- bung
- salt water
- three pipettes
- spatula
- permanent marker pen

Method

You need to decide what to put in each of your test-tubes to see what causes rusting to occur. Here are some points to consider:

- Boiled water will contain very little oxygen.
- A layer of oil prevents oxygen entering water.
- Anhydrous calcium chloride removes water from the air.
- Salt water affects the rate of rusting.

1   Fill in the table below to show what you will put in each tube and which of the following three variables each tube will contain
    - Oxygen
    - Water
    - Salt

Tube number	1	2	3	4
contents				
variable present				

2   Now that you have decided what each tube will contain, you need to write a numbered method for the investigation. Remember that you will need to label each tube.

....
....
....
....
....
....
....
....
....

Safety considerations

Wear eye protection throughout. The iron nails may be sharp so take care with these. Anhydrous calcium chloride is an irritant.

Recording data

Design a table to record your results.

Analysis

3 Which nail rusted the most?

....

....

4 Which nails rusted least?

....

....

5 Explain why you think that this was.

....

....

6 Based on the results of your experiment, which factors increase rusting?

....

....

Evaluation

7 What could you have done as a control for this investigation?

....

....

8 Suggest what would happen if you were to repeat this investigation using copper nails. Explain your answer.

....

....

....

....

Practical investigation 9.2 Preventing rusting

Objective

To protect iron and steel from rusting, a number of different methods have been developed. In this investigation, you will use a variety of these methods. Rusting is usually a slow process that can take from a few days to a few weeks to take place. To see more quickly if rusting is happening, an indicator solution will be used. This indicator changes colour from yellow to blue if rusting is taking place. By the end of this investigation you should be able to describe and explain methods of rust prevention. You may also be able to describe and explain the use of sacrificial protection as a method of rust prevention.

Equipment

- Magnesium ribbon
- copper foil
- eight test-tubes
- six iron nails
- correction fluid/acrylic paint
- test-tube rack
- corrosion indicator
- grease/petroleum jelly
- plastic wrap
- permanent marker pen
- galvanised nail (an iron nail coated in zinc)

Method

1. Place the eight test-tubes in the test-tube rack. Label them 1–8 with the permanent marker.
2. Paint one of the iron nails with the correction fluid or acrylic paint and set it to one side so that it can dry.
3. Leave test-tube 1 empty and place an iron nail in the test-tube labelled 2.
4. Cover one of the other iron nails in the grease/petroleum jelly. Place this in test-tube 3.
5. Wrap a piece of magnesium ribbon around an iron nail and place it in the test-tube labelled 4.
6. For tube 5 wrap a piece of copper foil around an iron nail and then add it to the tube.
7. Carefully wrap an iron nail in the plastic wrap. Make sure that there are no holes. Place this nail in test-tube 6.
8. Add the painted iron nail to test-tube 7.
9. Place the galvanized nail in test-tube 8.
10. Add the corrosion indicator to each of the tubes so that the nail is completely covered.
11. Leave for 30 minutes and then record your observations in the results table.

Safety considerations

Wear eye protection throughout.

Recording data

Design a results table to record your observations. Include a column for the colour change of the indicator.

Handling data

Fill your results in the following table.

Tubes where corrosion occurred	Tubes where no corrosion occurred

Analysis

1. In which tubes was corrosion prevented by stopping oxygen from coming into contact with the iron?

   ....

   ....

2. Why was magnesium ribbon better at preventing corrosion than copper foil?

   ....

   ....

3. Give an example of how each method of rust prevention is used in real life.
   a Plastic coating

   ....

b Painting

....

c Grease

....

d Galvanising

....

e Sacrificial protection

....

Evaluation

4 Why was one tube filled with just the corrosion indicator?

....

....

5 If the test-tube with just the corrosion indicator also changed colour, what would this tell us about the reliability of our investigation?

....

....

6 Suggest why there may still be some corrosion even in the tubes with nails that have been treated with grease, paint or plastic?

....

....

7 Electroplating is often used to prevent rusting. Explain how you could use electroplating to protect an iron nail from corrosion.

....

....

....

....

Practical investigation 9.3 Using limestone (calcium carbonate) to neutralise acidic water

Objective

Acidic lakes and rivers are a growing problem around the world. Acidic water can be caused by acid rain or chemicals used in farming and industry. If water becomes too acidic, plants and animals will no longer be able to live in it. In this investigation, you will process limestone to make lime (quicklime) and then use this to neutralise some acidic river water. By the end of this investigation you should be able to describe the manufacture of lime (quicklime) from limestone in terms of thermal decomposition and name some uses of lime (quicklime) and slaked lime.

Equipment

- Limestone
- Bunsen burner
- heat-resistant mat
- tin lid/pie tin
- beaker (100 cm³)
- measuring cylinder (50 cm³ or 100 cm³)
- tripod
- gauze
- glass rod
- spatula
- spotting tile (with 12 wells)
- Universal Indicator solution
- Universal Indicator colour chart
- distilled water
- river water
- pipette
- tongs
- two boiling tubes
- test-tube rack
- filter paper, filter funnel

Method

1. Set up the Bunsen burner, heat-resistant mat, tripod and gauze.
2. Place the tin lid on the gauze and add a few pieces of limestone. Ensure that the pieces are close together.
3. Light the Bunsen burner and heat the limestone pieces on a roaring blue flame directly for 5–10 minutes (Figure 9.1). This will turn the limestone into lime (quicklime).

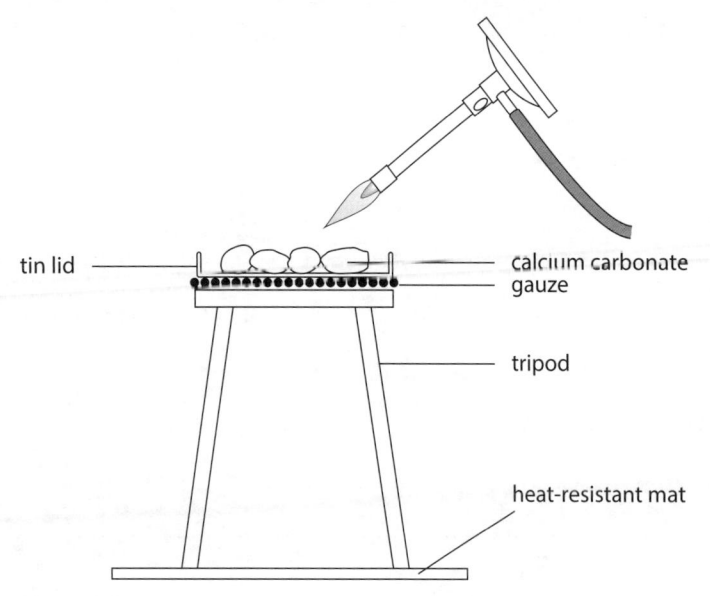

Figure 9.1

4. Allow the lime pieces to cool for a few minutes. While the pieces are cooling, add three drops of Universal Indicator solution into each of the wells on the spotting tile.
5. Then, using tongs, transfer the limestone pieces to a boiling tube.

6 Slowly add distilled water to the boiling tube using the pipette until the boiling tube is a quarter full. Record your observations on any reactions. Use the glass rod to stir the contents. You have now made slaked lime.
7 Fold the filter paper and make a cone shape. Place the paper in the funnel then filter the contents of the boiling tube into a clean, dry tube.
8 Using the measuring cylinder, pour 20 cm³ of river water into the beaker.
9 Using a pipette, take a small drop of the river water and add it to the first spot on the tile. Record the results.
10 Using a clean pipette, transfer 0.5 cm³ of your slaked lime solution into the beaker. Stir with the glass rod.
11 Use a pipette to take a few drops from the beaker and place them in the next spot on the tile. Record the results in your table.
12 Repeat steps **9–11** for the rest of the wells on your spotting tiles.

Safety considerations

Wear eye protection throughout.

Recording data

Write down your observations when adding water to the lime (quicklime).

....

....

Record the colour you observed for each tile in the results table below.

Tile	Volume of slaked lime added / cm³	Colour	pH
1	0.0		
2	0.5		
3	1.0		
4	1.5		
5	2.0		
6	2.5		
7	3.0		
8	3.5		
9	4.0		
10	4.5		
11	5.0		
12	5.5		

Analysis

1 What was the pH of the river water?

....
....

2 From your data, what volume of slaked lime was required to neutralise the river water?

....
....

3 Explain the reason for the colour changes you saw.

....
....
....
....
....

Evaluation

4 Suggest an improvement to the method that would have allowed you to get more accurate data?

....
....

5 What technique or apparatus could you have used to measure a more precise volume of the slaked lime needed for the neutralisation of the river water?

....
....

6 If some scientists wanted to neutralise an acidic lake, what factors would they need to consider?

....
....
....
....
....
....

Exam-style questions

1 Thomas and Christine are investigating rusting (see Figure 9.2). They have set up the experiment below looking at how different variables affect whether an iron nail will rust or not.

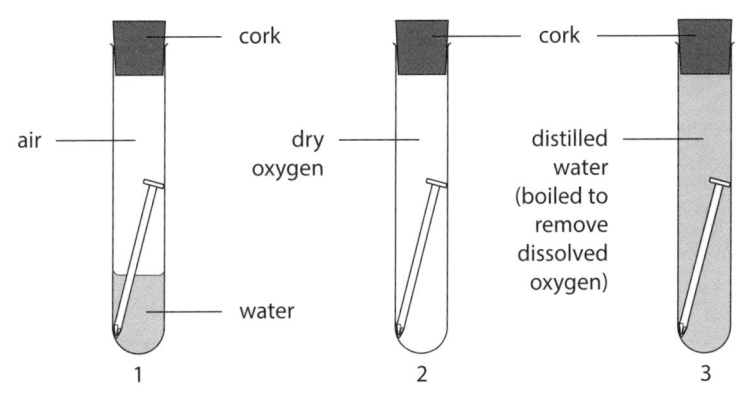

Tube 1: air, water
Tube 2: dry oxygen
Tube 3: distilled water (boiled to remove dissolved oxygen)

Figure 9.2

a Predict which of the tubes rust will appear in first. [1]

....

b Explain your prediction. [2]

....

....

c Predict which of the tubes rust will appear in last. [1]

....

d Explain your prediction. [2]

....

....

In another investigation, Sherif and Jing are investigating sacrificial protection of iron. They have five metal strips which they will wrap around an iron nail. The nail will then be placed in water that contains oxygen.

e In the box below, suggest which of the metals would protect the iron nail from rusting and which would not. [5]

Metals: zinc, copper, tin, magnesium, lead

Metals that would protect an iron nail from rusting	Metals that would not protect an iron nail from rusting

9 Industrial inorganic chemistry

f Explain the reasons for your predictions above. [2]

....

....

Total [13]

2 Sinead and Noshaba are making slaked lime from calcium carbonate and water. They used the apparatus shown in Figure 9.3 to first heat the calcium carbonate and then add water. The solution was then filtered.

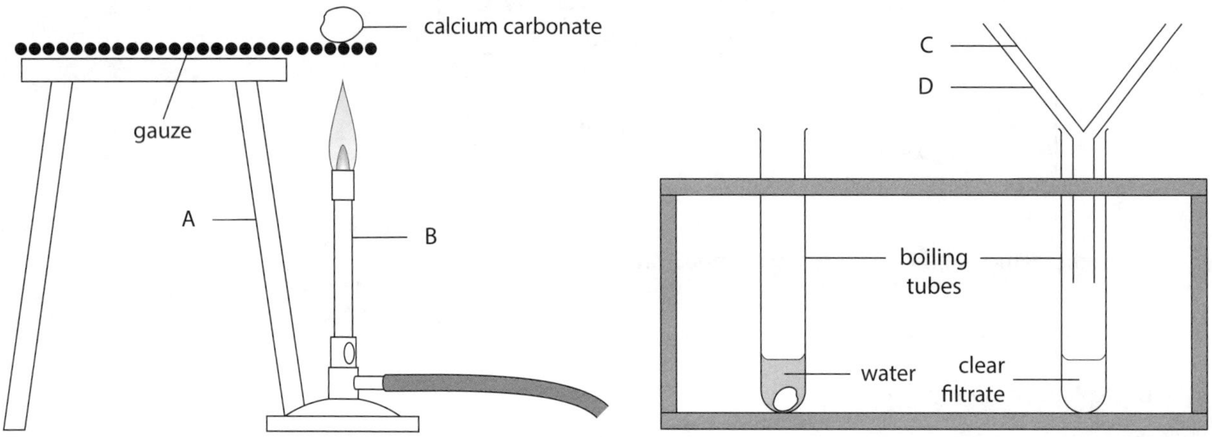

Figure 9.3

a State the names of the different pieces of apparatus. [4]

A ....

B ....

C ....

D ....

b Name the type of reaction that occurs when calcium carbonate is heated. [1]

....

c Write the word equation for the reaction that takes place as calcium carbonate is heated. [2]

....

d Write the symbol equation for the reaction that takes place as calcium carbonate is heated. [2]

....

After the calcium carbonate was heated, it was placed in a boiling tube. Water was added slowly using a pipette. Lots of steam was produced.

e What type of reaction was taking place as the water was added? [1]

....

Total [10]

3 Car engines become very hot when they are being used and so a cooling system is used. Part of the system is a radiator which transfers heat to the surroundings. The coolant usually contains water and radiators are normally made of steel and so they are prone to rusting. Some coolant additives are sold that can be added to the radiator to reduce the rusting that takes place. You have been given four different types of coolant additive to test. Small strips of the metal used to make the steel radiator have also been provided. Plan an investigation to determine which of the coolant additives works best at preventing rusting. [10]

....
....
....
....
....
....
....
....
....
....

Total [10]

10 Organic chemistry

In this chapter, you will complete investigations on:

◆ **10.1** Testing for alkanes and alkenes

◆ **10.2** Fermentation of glucose using yeast

◆ **10.3** Making esters from alcohols and acids

Practical investigation 10.1 Testing for alkanes and alkenes

Objective

Hydrocarbons can be separated into two groups: **alkanes** and **alkenes**. The difference between them is that alkanes are saturated, which means they contain no carbon–carbon double bonds. Alkenes are unsaturated, which means that they contain at least one carbon–carbon double bond. There are two very simple tests to find out whether a hydrocarbon is saturated or not. One is aqueous bromine, which decolourises from orange-brown in the presence of an alkene. Though not part of the syllabus, another indicator that can be used is an acidified dilute solution of potassium manganate(VII), which also decolourises, from purple to colourless. By the end of this investigation you should be able to distinguish between saturated and unsaturated hydrocarbons using aqueous bromine.

Equipment

- Test-tube rack
- pipettes
- eight test-tubes
- eight rubber bungs
- aqueous bromine
- dilute acidified potassium manganate(VII)
- hexane
- hexene
- unknown hydrocarbons samples A and B

Method

Aqueous bromine test
1. Add 2 cm³ of hexane to one of the test-tubes.
2. Using a pipette, slowly add six drops of aqueous bromine into the test-tube. Close the test-tube with the rubber bung and then shake it to mix the two liquids together. Record the colour in the table.
3. Repeat steps **1–2** with the other hydrocarbon samples.

Dilute acidified potassium manganate(VII) test
1. Add 2 cm³ of hexane to one of the test-tubes.
2. Using a pipette, slowly add six drops of dilute acidified potassium manganate(VII) into the test-tube. Close the test-tube with the rubber bung and then shake it to mix the two liquids together. Record the colour in the table.
3. Repeat steps **1–2** with the other hydrocarbon samples.

Safety considerations

Wear eye protection throughout. Aqueous bromine is corrosive so wear gloves when handling it. All of the hydrocarbon samples are flammable. Acidified potassium manganate(VII) is harmful and can stain hands and clothing.

Recording data

Design a results table to record your data.

Analysis

1 Write your results in the table below.

Saturated hydrocarbons	Unsaturated hydrocarbons

2 Products made from animal fats usually contain saturated hydrocarbon chains while products from plants tend to contain unsaturated hydrocarbon chains. Predict the results for each type of product in the table below.

Product	Test result with aqueous bromine	Test result with dilute acidified potassium manganate(VII)
milk		
olive oil		
sesame oil		
cream		

Evaluation

3 Why was it useful to conduct more than one test for alkenes?

....

....

Practical investigation 10.2 Fermentation of glucose using yeast

Objective

Ethanol is an important chemical used in medicine and industry. The majority of the ethanol we use is produced by the fermentation of sugars using yeast, although it is also possible to produce ethanol by adding steam to ethene in the presence of a catalyst. The process of fermentation is also used to produce bread and alcoholic drinks and involves yeast undergoing anaerobic respiration. This means that the yeast respires without using oxygen. By the end of this investigation you should be able to describe the manufacture of ethanol by fermentation.

$$\text{glucose} \xrightarrow{\text{yeast}} \text{ethanol} + \text{carbon dioxide}$$

Equipment

- Boiling tube
- spatula
- glucose solution
- yeast
- oil
- measuring cylinder (50 cm³)
- bung and delivery tube
- two test-tubes
- water bath/beaker of warm water (40 °C)
- test-tube rack
- test-tube holder
- limewater

Method

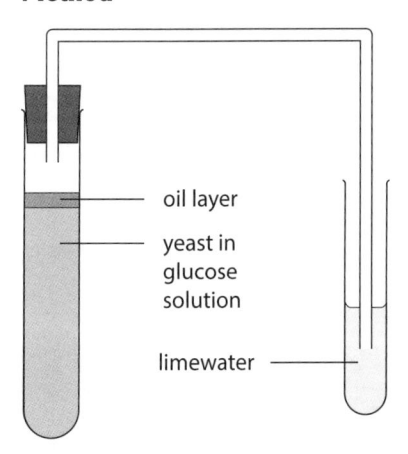

Figure 10.1

1. Add 20 cm³ of glucose solution to the boiling tube. Add two heaped spatulas of yeast to the tube and close using the bung. Carefully add 0.5 cm³ of oil to the boiling tube so that it forms a thin layer on top of the mixture (Figure 10.1).
2. Half fill two test-tubes with limewater and place them in the test-tube rack.
3. Place the boiling tube into the water bath or beaker of warm water. Place the end of the delivery tube so that it is submerged in one of the test-tubes with limewater. Bubbles should start to come out of the delivery tube.
4. After 40 minutes, observe the two tubes with limewater and record your results.

Safety considerations

Wear eye protection when using the limewater.

Recording data

Design a table to record your results.

Analysis

1   Complete the following passage using the missing words below.

> respiration   oxygen   bread   yeast   carbon dioxide   glucose

Fermentation is a process where .... is turned into ethanol and .... by the microorganism called .... This occurs when yeast undergo anaerobic .... This is respiration in the absence of .... Fermentation is used in the production of .... and alcoholic drinks.

2   What do the limewater results show the presence of?

....

Evaluation

3   What was the purpose of the test-tube of limewater that did not have the delivery tube in it?

....

....

4   Why was a layer of oil added to the mixture?

....

5   Research how you could test the solution for the presence of ethanol or an alcohol.

....

....

Practical investigation 10.3 Making esters from alcohols and acids

Objective

Esters are a family of compounds that have strong and often pleasant smells. They are naturally occurring in both fruits and flowers. We use them in food flavourings and also in perfumes. In this practical, you will make a variety of different esters. By the end of this investigation you should be able to describe the reaction of a carboxylic acid with an alcohol in the presence of a catalyst to give an ester.

Equipment

- Six test-tubes
- pipettes, beaker (250 cm³)
- test-tube rack
- Bunsen burner
- heat-resistant mat
- tripod and gauze
- test-tube holder
- measuring cylinder (50 cm³)
- timer/stopwatch
- sodium carbonate solution
- concentrated ethanoic acid
- propanoic acid
- concentrated sulfuric acid
- three alcohols from: methanol, ethanol, propan-1-ol, butan-1-ol, butan-2-ol, propan-2-ol

Method

You will need to decide on three different combinations of acids (ethanoic or propanoic) and alcohols (methanol, ethanol, propan-1-ol, propan-2-ol, butan-1-ol and butan-2-ol). Add these into the results table now.

1. Add one drop of concentrated sulfuric acid to each of four test-tubes (if this has not already been done for you).
2. Add 8–10 drops of ethanoic acid or propanoic acid to the sulfuric acid in the first test-tube.
3. Using a pipette, add 8–10 drops of your chosen alcohol to the first test-tube.
4. Use the measuring cylinder to add 50 cm³ of water to the 250 cm³ beaker. Carefully lower the tube into the beaker so that it stands upright but at a slight slant.
5. Set up the Bunsen burner, tripod and gauze on the heat-resistant mat well away from your test-tube.
6. On a cool blue flame, heat the beaker gauze until the water begins to boil. Once the water is boiling, stop heating immediately. Turn the Bunsen burner off.
7. Let the test-tube rest in the hot water for approximately 1 minute. Observe the mixture in the test-tube carefully. If it begins to boil, use the test-tube holder to lift the test-tube out of the water. Wait until the mixture stops boiling, and then return it to the beaker.
8. After 1 minute has passed, remove the test-tube and place it in the test-tube rack. Allow the test-tube to cool.
9. Half fill a test-tube with the sodium bicarbonate solution.
10. Slowly pour the cooled mixture into the sodium bicarbonate solution. Pour the liquids from tube to tube a few times to ensure that they are well mixed.
11. Once you stop mixing, you should be able to see a layer forming on the top – this is the ester.
12. Use your hand to waft the odour from the mouth of the test-tube towards your nose. Record your observations in the results table.
13. Repeat steps **2–12** with each combination of acid and alcohol you have chosen to use.

Safety considerations

Wear eye protection throughout. Concentrated ethanoic acid, propanoic acid and concentrated sulfuric acid are corrosive. Methanol is toxic and highly flammable. Ethanol is harmful and highly flammable. Propan-1-ol and propan-2-ol are irritants and both highly flammable. Butan-1-ol is harmful and butan-2-ol is an irritant. Make sure that you keep the flammable alcohols away from the Bunsen flame.

10 Organic chemistry

Recording data

Test-tube	Acid used	Alcohol used	Smell produced
1			
2			
3			

Analysis

The first word in the name of an ester comes from the alcohol used.

Alcohol name	Ester name first word
methanol	methyl
ethanol	ethyl
propanol	propyl
butanol	butyl

The second word that forms the name of the eater comes from the acid used.

Acid name	Ester name second word
ethanoic	ethanoate
propanoic	propanoate

1 Using the information in the tables, suggest the names of the esters that you made.
   a Test-tube 1 ....
   b Test-tube 2 ....
   c Test-tube 3 ....

2 Which alcohol and acid combination would need to be used to make the following ester: butyl ethanoate?

   ....

Evaluation

3 It is very likely that the first step of this experiment will have been done for you. Suggest why this might have been.

   ....
   ....

152 Cambridge IGCSE Chemistry

Exam-style questions

1 Stewart and Deirdre were using yeast to ferment glucose.

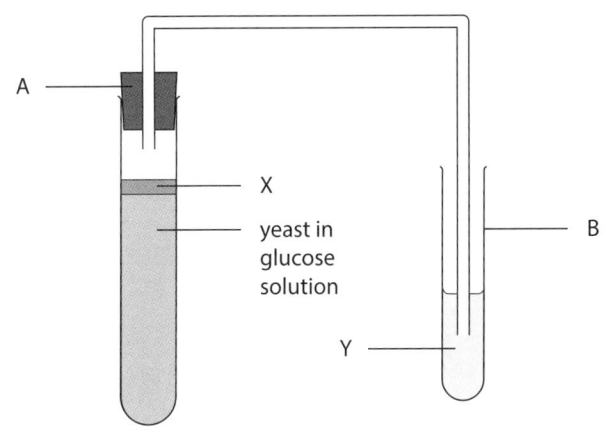

Figure 10.2

a Identify the apparatus labelled in Figure 10.2. [2]
A ....
B ....

b Identify liquid X and suggest the purpose of it in this experiment. [2]

....

....

....

c Identify liquid Y and suggest the purpose of it in this experiment. [2]

....

....

....

d The students heated the boiling tube with the yeast and glucose solution to 80 °C for 3 minutes. Predict and explain the result you would expect to see in the liquid marked Y. [4]

....

....

....

Total [10]

2 As part of a healthy lifestyle, people are advised to cut down on saturated fats and increase their intake of unsaturated fats. You are given five different food spreads and you must design an investigation that would enable you to determine which ones contain unsaturated hydrocarbons. Remember to include safety precautions. [7]

....
....
....
....
....
....
....
....
....
....

Total [7]

11 Petrochemicals and polymers

> **In this chapter, you will complete investigations on:**
>
> ◆ 11.1 Cracking hydrocarbons
>
> ◆ 11.2 Comparing polymers
>
> ◆ 11.3 Comparing fuels

Practical investigation 11.1 Cracking hydrocarbons

Objective

Hydrocarbons are made of hydrogen and carbon atoms, and they come in two varieties: saturated and unsaturated. Saturated hydrocarbons contain no carbon–carbon double bonds while unsaturated hydrocarbons contain one or more carbon–carbon double bond. The composition of crude oil is not the same everywhere that it is found and it usually contains more long-chain hydrocarbons than short-chain ones. Generally, short-chain hydrocarbons are more useful and more valuable than long-chain ones. It is possible to convert long-chain hydrocarbons into short-chain hydrocarbons (including alkenes) and hydrogen using a process called **cracking**. By the end of this investigation you should be able to describe the manufacture of alkenes and of hydrogen by cracking.

Equipment

- Three test-tubes with rubber bungs
- test-tube rack
- boiling tube
- bung with hole to fit boiling tube and delivery tube
- clamp stand with boss and clamp
- water trough or large beaker
- delivery tube with valve fitted
- Bunsen burner
- heat-resistant mat
- liquid paraffin
- porcelain crushed into fragments
- mineral wool
- aqueous bromine
- wooden splint
- pipette
- glass rod

Method

1. Add mineral wool to the boiling tube and press it down to the bottom using the glass rod. It needs to be between 2 cm and 3 cm depth from the bottom of the tube.
2. Using the pipette, add 2 cm³ of liquid paraffin to the mineral wool.
3. Clamp the boiling tube near the mouth of the tube and then rotate the tube so that it is horizontal. Tighten the boss so that the tube is held securely in place (see Figure 11.1).

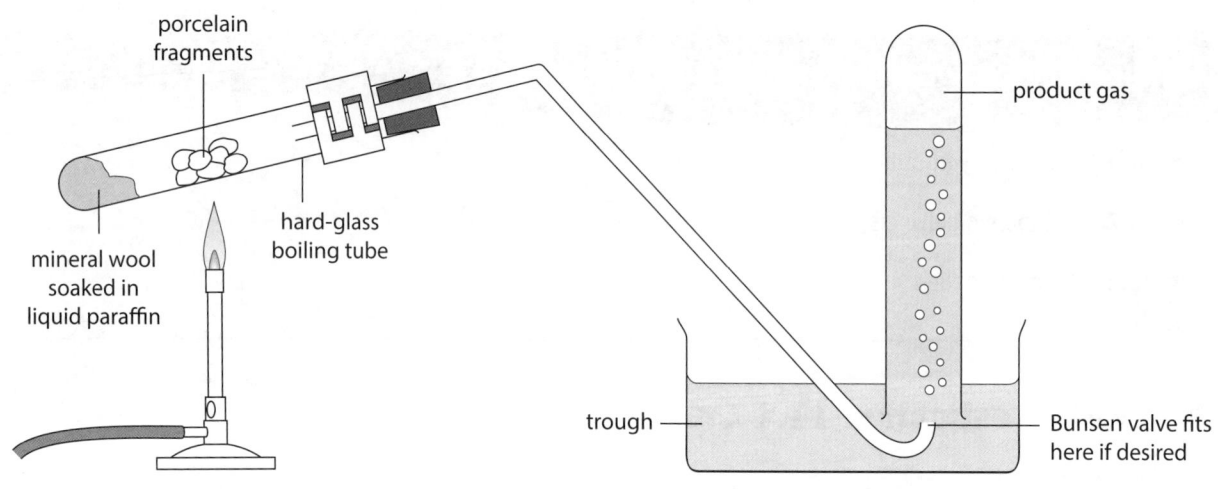

Figure 11.1

4. Carefully add some of the crushed porcelain fragments to the tube and push them to the middle of the tube using the glass rod. Loosen the boss and gently rotate the boiling tube so that there is a slight downward tilt towards the mineral wool. Retighten the boss.
5. Attach the delivery tube to the rubber bung with the hole and then place the bung securely on the mouth of the boiling tube. Make sure that there is a valve fitted (see Figure 11.2) to the delivery tube.

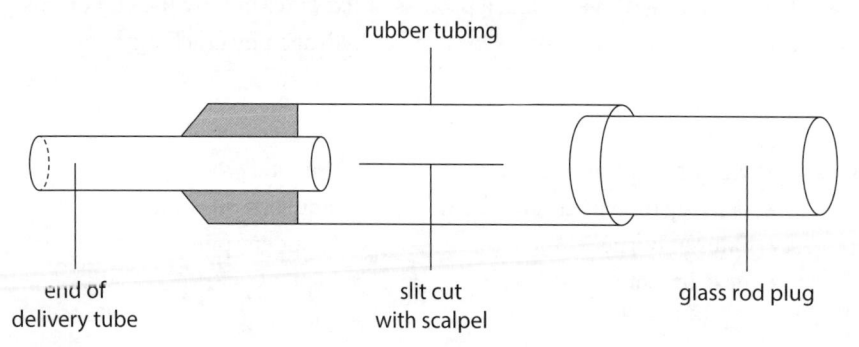

Figure 11.2

6. Half fill the trough or beaker with water. Fill the test-tubes with water and then place your finger over the end of the tube. Invert the tubes and then place them in the beaker. Add the test-tube bungs to the trough or beaker so that you can easily push the test-tubes down on to them.
7. Place the delivery tube in the water trough or beaker so that it is well submerged. When you heat the boiling tube, gas will come out of the valves and it will be collected in the inverted test-tubes.
8. Light the Bunsen burner and adjust it to give a gentle blue flame. Heat the porcelain fragments until they begin to glow slightly red. At this point you can start to heat the mineral wool. You will need to move the Bunsen burner back and forth between the porcelain fragments and mineral wool. Avoid heating the rubber bung.

9 Bubbles will start to come from the valve in the water. Use the test-tube to collect the gas. Once the tube is full of gas, close it with a bung from the bottom of the beaker/trough and place it in the test-tube rack.
10 Once all of the tubes are full, lift the delivery tube out of the water. Only at this point will it be safe to stop heating the boiling tube.
11 Open the first test-tube and carefully smell the contents. Record your observations in the results table.
12 Use the pipette to add a few drops of aqueous bromine to the second tube. Shake the tube and record your observations.
13 Light a wooden split and open the third tube. Place the wooden splint near the mouth of the tube. Record your observations.

Safety considerations

Wear eye protection throughout. Be very careful with the hot glassware. Make sure that you keep heating the boiling tube – if you stop, there is the possibility that water will be sucked back into the delivery tube. This is very dangerous and can cause the tube to shatter. If you see water being sucked back into the tube, loosen the boss and lift the whole apparatus up higher so that the delivery tube is no longer in the water. Aqueous bromine is harmful.

Recording data

Complete the results table.

Tube	Observation
1 smell	
2 aqueous bromine	
3 lighted splint	

Analysis

1 With reference to your results, draw a conclusion about the gas in the test-tubes.

....

....

Evaluation

2 Gas was produced when the porcelain fragments were being heated. This was before the mineral wool was heated. Why were bubbles coming out of the valve?

....

....

3 Why was it necessary to heat the boiling tube for the duration of the experiment?

....

....

4 Why were the porcelain fragments used?

....

5 Why were the porcelain fragments heated?

....

Practical investigation 11.2 Comparing polymers

Objective

There are many different types of plastic and it can be difficult to identify them. Recycling is becoming increasingly important and it is sometimes necessary to identify plastics that are unlabelled. Different plastics have different densities and it is possible to determine the identity of an unknown sample by using liquids of known density. By the end of this investigation you should be able to use the differences in their densities to identify different plastics.

Equipment

- Five boiling tubes
- test-tube rack
- permanent marker
- glass rod
- samples of four different plastics (five pieces of each)
- liquids 1–5, measuring cylinder (50 cm^3 or 100 cm^3)

Method

1. Label the boiling tubes 1–5 and place them in the test-tube rack.
2. Add 20 cm^3 of one of the five liquids to the matching-numbered boiling tube. Remember to wash the measuring cylinder between each liquid.
3. Add a sample of each plastic to each of the tubes.
4. Using the glass rod, stir the liquid in each boiling tube. Remember to wash the glass rod after stirring each tube to avoid contaminating the liquids.
5. Look at each boiling tube and record whether the samples of plastic float or sink. Record the data in the table below.

Safety considerations

Wear eye protection throughout. Ethanol (which will be in tubes 1 and 2) is flammable. Potassium carbonate (in tubes 4 and 5) is an irritant.

Recording data

For each sample, record whether it sank or floated in each liquid by writing an S (sank) or an F (floated).

Sample	Tube				
	1	2	3	4	5
1					
2					
3					
4					

Handling data

Based on the results and the data given in Tables 11.1 and 11.2, determine the density range of each polymer,

Liquid	Density / g/cm³
1	0.91
2	0.94
3	1.00
4	1.15
5	1.38

Table 11.1 Liquid density chart

Polymer name	Density range / g/cm³
polypropylene	0.89–0.91
low-density polyethylene	0.91–0.93
high-density polyethylene	0.94–0.96
polystyrene	1.04–1.12
polyvinyl chloride (PVC)	1.20–1.55
polyethylene terephthalate	1.38–1.41

Table 11.2 Polymer density chart

Sample	Density range / g/cm³
1	
2	
3	
4	

Analysis

1 Complete the following sentences to identify each type of polymer:

Sample 1 was .... because it was more dense than liquid .... but less dense than liquid ....

Sample 2 was .... because it was more dense than liquid .... but less dense than liquid ....

Sample 3 was .... because it was more dense than liquid .... but less dense than liquid ....

Sample 4 was .... because it was more dense than liquid .... but less dense than liquid ....

Evaluation

2 Why might there be some error in your results?

....

....

3 Polymer X has a density of 0.98 g/cm³. Predict the results you would obtain if it was placed into the same five liquids used in your investigation. Write an S for sink and an F for float.

Tube	1	2	3	4	5
X					

Practical investigation 11.3 Comparing fuels

Objective

Crude oil is a fossil fuel which causes pollution when combusted and is also a non-renewable resource. This means that it will eventually run out and so alternative fuels will need to be used. One alternative to fossil fuels is ethanol, which can be made by fermenting sugar made by plants. As it is made using plants, ethanol is considered to be a renewable resource. The plants absorb carbon dioxide from the atmosphere when they grow and the ethanol produces less harmful pollution when it is burnt than fossil fuels do. By the end of this investigation you should be able to name the uses of some fractions of petroleum.

Equipment

- Clamp and stand
- heat-resistant mat
- thermometer
- measuring cylinder (100 cm^3)
- mineral wool/ceramic wool
- pipette
- boiling tube
- crucible
- paraffin
- ethanol

Method

You need to design a method to compare the two fuels.

My independent variable is ....

My dependent variable is ....

My control variables will be ...., ..., ...

I will make my results accurate by:

....

....

I will make results reliable by:

....

....

This is how I will set up my apparatus:

To make sure I work safely, I will:

....

....

My method will be:

....

....

....

....

....

....

....

....

....

Once you have written your method get your teacher to check that it is safe before you begin the experiment.

Safety considerations

Wear eye protection at all times. Ethanol and paraffin are flammable.

Recording data

Design a results table to record your results.

Analysis

1 Write a conclusion for your experiment to explain which of the two fuels was better. Make sure that you include references to the actual experimental data.

....
....
....
....
....

Evaluation

2 Only some of the heat energy from the burning fuel was used to heat the water. Where did the rest of it go?

....
....

3 How could you have improved your experiment so that less of this heat was lost?

....
....

4 Can you explain why many people are reluctant to stop using fossil fuels even when there are alternatives available?

....
....
....
....

5 Imagine you are trying to persuade someone to switch to using biofuels. What arguments would you use?

....
....
....
....
....

Exam-style questions

1 Malak and Praise were trying to crack a sample of long-chain hydrocarbons into shorter-chain hydrocarbons.

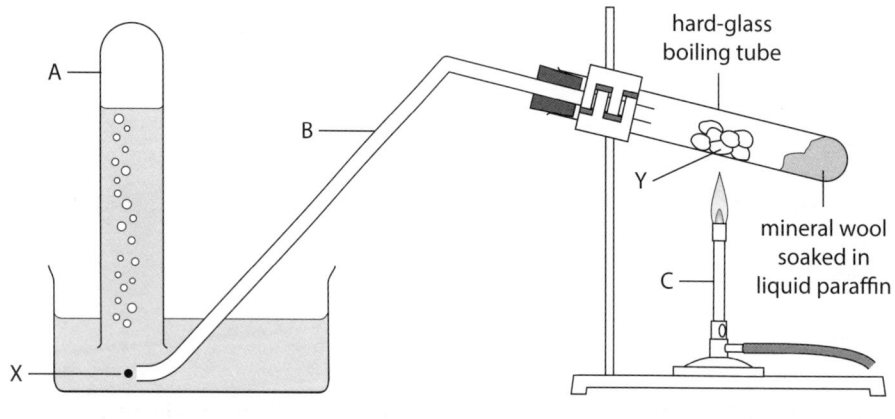

Figure 11.3

a Identify the apparatus in the diagram. [3]
A ....
B ....
C ....

b Name the part of the diagram labelled Y and suggest its purpose. [2]

....

....

c This experiment has not been set up safely. With specific reference to the part of the diagram labelled X, explain why this is a dangerous way to complete this practical. [2]

....

....

d Describe how you could test for the presence of alkenes in the gas produced by cracking. [2]

....

....

Total [9]

2 Samirah and Ben were comparing two fuels to see which one was better. They set up the apparatus up below. They were comparing methanol and paraffin.

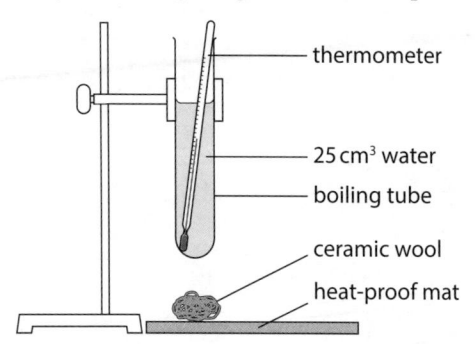

Figure 11.4

164 Cambridge IGCSE Chemistry

The two students used the methanol first. They measured 1 cm³ of the fuel and placed it in the crucible. They then added 25 cm³ of water to the boiling tube. They measured the temperature of the water and then lit the fuel. After the fuel had finished burning, they measured the temperature again.

a Look at the thermometer readings in the table. For each repeat for methanol, calculate the temperature change. [3]

Repeat number	Thermometer at start / °C	Thermometer at finish / °C	Temperature change / °C
1	23	28	
2	23	37	
3	22	27	

The pair then did the experiment again but this time they used paraffin.

b Look at the thermometer readings in the table. For each repeat for paraffin, calculate the temperature change. [3]

Repeat number	Thermometer at start / °C	Thermometer at finish / °C	Temperature change / °C
1	23	53	
2	23	43	
3	22	53	

c Calculate the mean average temperature change for each fuel. [2]
Average temperature change for methanol = ... °C
Average temperature change for paraffin = ... °C

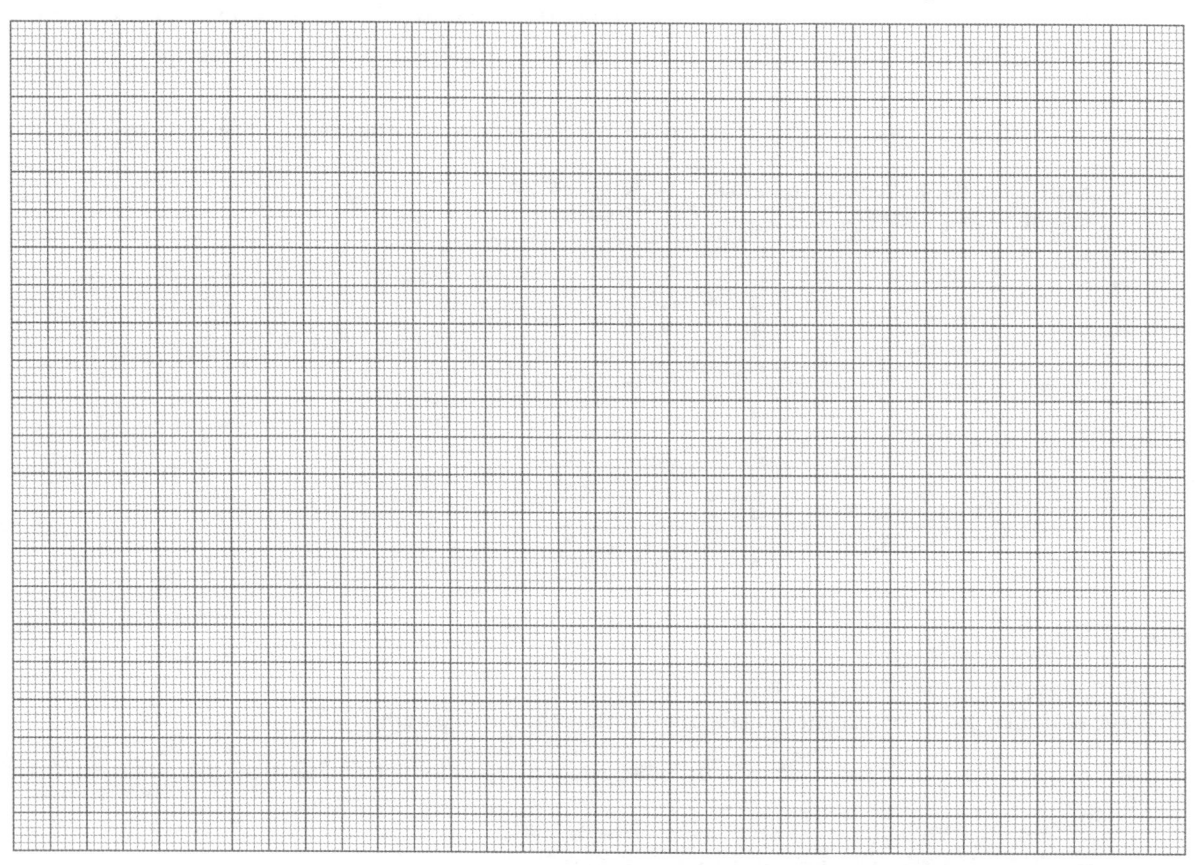

d Plot a bar graph to compare the average temperature change for the two fuels on the grid below. [4]

Total [12]

12 Chemical analysis and investigation

In this chapter, you will complete investigations on:

♦ 12.1 Identifying anions

♦ 12.2 Identifying cations

Practical investigation 12.1 Identifying anions

Objective

In chemistry, it is often important to be able to identify an unknown substance. One way in which it is possible to do this is to use qualitative analysis. This is a series of reactions that can be used to detect the presence or absence of an ion. In this practical, you will be testing for different **anions** (negative ions). It is important that you make careful observations and record them accurately. By the end of this investigation you should be able to describe the tests and results for anions.

Equipment

- Test-tube rack
- eight test-tubes
- pipettes
- nitric acid
- Bunsen burner
- heat-resistant mat
- aluminium foil
- red litmus paper

- rubber bung and delivery tube
- spatula
- calcium carbonate
- potassium sulfate
- potassium iodide
- sodium bromide
- limewater
- barium nitrate solution

- silver nitrate solution
- ammonia solution (1.0 mol/dm^3)
- copper chloride
- potassium sulfite
- copper(II) nitrate(V)
- dilute hydrochloric acid
- sodium hydroxide solution
- potassium manganate(VII) solution

Method

Tests can be completed in any order.

Testing for carbonates
1. Add two spatulas of calcium carbonate to a clean test-tube. Half fill another test-tube with limewater. Take the delivery tube and place the end so that it is submerged in the limewater.
2. Using a pipette, add approximately 2 cm³ of nitric acid to the calcium carbonate. Immediately close the tube using the rubber bung. Observe the reaction in both tubes and record your results in the table in the results section.

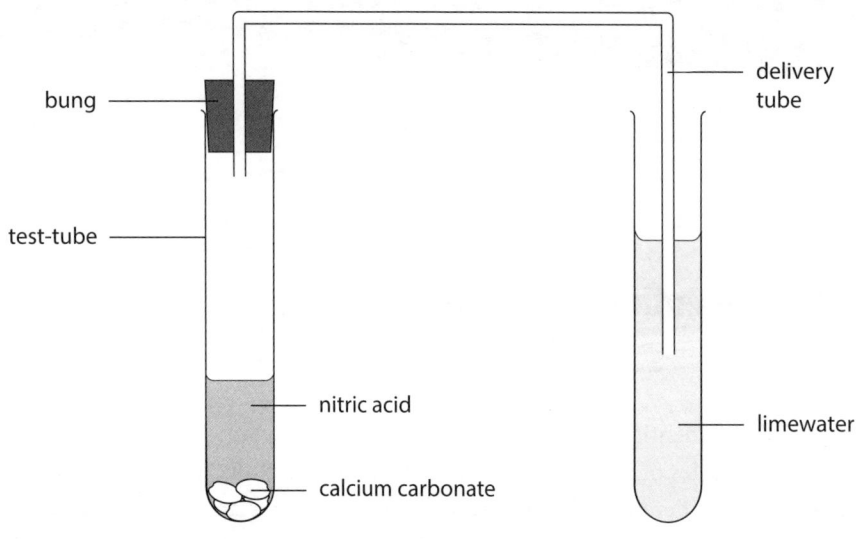

Figure 12.1

Testing for chlorides, bromides and iodides
1. Add two spatulas of copper chloride to a clean test-tube.
2. Half fill the test-tube with nitric acid.
3. Gently shake the tube so that the solid dissolves.
4. Using a pipette, add 2 cm³ of silver nitrate solution to the test-tube. In the results table, record the colour of the precipitate formed.
5. Using a pipette slowly add a few drops of ammonia solution. Record whether the precipitate dissolves or not.
6. Repeat steps **1–5** with sodium bromide and then potassium iodide instead of copper chloride.

Testing for sulfates
1. Add two spatulas of potassium sulfate to a clean test-tube.
2. Half fill the test-tube with nitric acid.
3. Gently shake the test-tube so that the solid dissolves.
4. Using a pipette, add 2 cm³ of barium nitrate solution to the test-tube. Record the colour of the precipitate formed in the results table.

Testing for sulfites
1. Add approximate 2 cm³ of potassium sulfite solution to a clean test-tube.
2. Using a pipette, add 2 cm³ of dilute hydrochloric acid to the test-tube.
3. Add a few drops of aqueous potassium manganate(VII) solution to the test-tube. Record the colour change in the results table.

Testing for nitrates
1. Fill a test-tube to about 3 cm depth with copper(II) nitrate(V) solution.
2. Add 2 cm³ of sodium hydroxide solution.
3. Place a small strip of aluminium foil in the test-tube.
4. Using the Bunsen burner, carefully heat the test-tube.
5. Hold a damp piece of red litmus paper over the mouth of the test-tube. Record the result in the results table.

Safety considerations

Wear eye protection throughout. Nitric acid is corrosive. Sodium hydroxide and silver nitrate are irritants. Copper chloride and barium nitrate are toxic. Both potassium manganate(VII) and copper(II) nitrate(V) solutions are harmful. Take care not to breathe in the fumes produced when heating the copper(II) nitrate(V) solution.

Recording data

When recording data it is important to remember the difference between an observation and a conclusion. An observation is something that you can physically experience (see, hear or feel). Examples of observations would be to see effervescence in a test-tube or hear a squeaky pop.

A conclusion is the explanation for the observation so you would see effervescence if a gas was being produced or hear a squeaky pop if the gas was hydrogen.

Ion being tested	Observations
carbonate	
chloride	
bromide	
iodide	
sulfate	
sulfite	
nitrate	

12 Chemical analysis and investigation

Analysis

1 What can you conclude from the results of the carbonate test about the identity of the gas produced?

....

....

2 What colour precipitate would you expect to be formed if sodium chloride were added to nitric acid and then silver nitrate solution were added?

....

....

3 What is the name of the salt formed during the sulfate test?

....

4 Using your knowledge of testing for gases, name the gas produced during the nitrate test.

....

Practical investigation 12.2 Identifying cations

Objective

Because salts are ionic, they will always contain both an anion and a cation. From carrying out Investigation 12.1 you will be confident at identifying anions. You must now learn how to identify **cations** (positive ions). This will mean that, by performing the tests you have learnt, you will be able to identify both the cation and anion of a salt. By the end of this investigation you should be able to describe the tests and results for cations.

Equipment

- Test-tube rack
- test-tubes
- pipettes
- Bunsen burner
- heat-resistant mat
- red litmus paper
- tongs
- solutions of: sodium hydroxide, ammonium ion solution, copper sulfate, iron(II) carbonate, iron(III) sulfate, chromium(III) potassium sulfate, calcium chloride, zinc sulfate, aluminium sulfate

Method

Test for ammonium ions

1. Add 2 cm^3 of ammonium ion solution and 2 cm^3 of sodium hydroxide to a test-tube. Set up the Bunsen burner on the heat-resistant mat. Heat the solution gently. Using the tongs, hold a piece of damp red litmus paper over the mouth of the tube (see Figure 12.2). Record your observations in the results table.

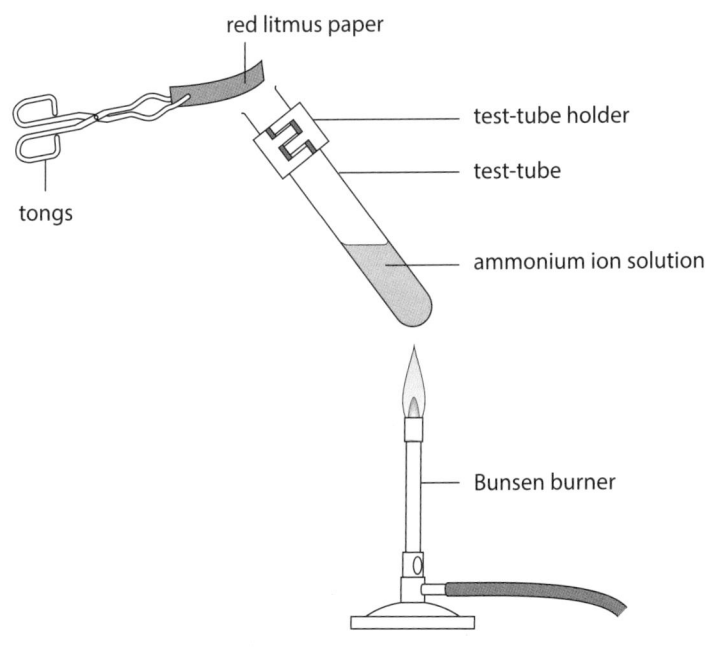

Figure 12.2

Test for metal cations

1. Add 2 cm3 of copper sulfate to a test-tube. Slowly (drop by drop) add 2 cm^3 of sodium hydroxide. Record your observations in the results table.
2. Add 2 cm^3 more sodium hydroxide solution. Record your observations in the results table.

3 Add 2 cm³ of copper sulfate to a clean test-tube. Slowly add 2 cm³ of ammonia solution. Record your observations in the results table.
4 Add 2 cm³ more ammonia solution. Record your observations in the results table.
5 Repeat with each of the different cation solutions.

Safety considerations

Wear eye protection throughout. Sodium hydroxide is corrosive. Ammonia solution is an irritant. Take care not to breathe in the fumes produced when heating the ammonium ion solution.

Recording data

Cation solution	Result when sodium hydroxide is added	Result when excess sodium hydroxide is added	Result when ammonia solution is added	Result when excess ammonia solution is added
ammonium hydroxide				
copper sulfate				
iron(II) carbonate				
iron(III) sulfate				
chromium(III) potassium sulfate				
calcium chloride				
zinc sulfate				
aluminium sulfate				

Handling data

In the space below, design a flow diagram that you could use to identify any of the metal ions in salts. The first question should be 'Is a precipitate formed when sodium hydroxide is added?'

Analysis

1 Look at your result for the damp red litmus paper test you did on the heated ammonium solution after you added sodium hydroxide. What does this result tell you about the identity of the gas produced?

....

....

2 Identify the following cation for the experimental results given below:
- Result when sodium hydroxide added = white precipitate (ppt) formed
- Result when excess sodium hydroxide added = ppt dissolves giving a clear solution
- Result when ammonia solution added = white ppt formed
- Result when excess ammonia solution added = ppt dissolves

The cations in the investigation are ....

Evaluation

3 Why is it important that clean test-tubes are used for each experiment?

....

Exam-style questions

1 Philippos is trying to identify an unknown salt. He first makes a visual observation of the salt. It is a white powder. He then adds dilute hydrochloric acid to the powder.

a What would Philippos expect to see if the powder contained a carbonate? [1]

....

From the results of the first test the student knows that the substance is not a carbonate.
He adds some dilute nitric acid to a sample of the powder and then some silver nitrate solution. A pale white or creamy precipitate is formed.

b From this result, which two anions might the sample contain? [2]

....

c What further test could Philippos do to identify which of the two anions identified in part **b** are in the sample? [2]

....

Philippos now wants to identify the cation present. He adds a small amount of sodium hydroxide and then warms the test-tube.

d How could he test for the presence of ammonia gas being given off? [1]

....

e If ammonia gas were given off, what would be the result of the test you described in **d**? [1]

....

174 Cambridge IGCSE Chemistry

No ammonia gas was given off but a white precipitate was formed. Excess sodium hydroxide was added but the precipitate did not dissolve.

Ammonia solution was added to a fresh sample of the powder. A white precipitate was formed that was soluble in excess ammonia.

f What was the cation present in the sample? [1]

....

Total [8]

2 Look at the results table below and write the results of each test you would expect for the salts listed. [9]

Salt	Sodium hydroxide added / and in excess	Ammonia solution added / and in excess	Dilute nitric acid added and silver nitrate added
calcium chloride			
zinc iodide			
chromium(III) bromide			

Total [9]

3 You have been given a bottle with a solution containing a salt with a magnesium cation but an unknown anion. Plan an investigation to explain how you would identify the anion present in the solution. [12]

....
....
....
....
....
....
....
....
....
....
....
....

Total [12]

12 Chemical analysis and investigation